Innovation Policy and the Limits of Laissez-faire

Innovation Policy and the Limits of Laissez-faire

Hong Kong's Policy in Comparative Perspective

Edited by

Douglas B. Fuller

palgrave
macmillan

First published 2010 by
PALGRAVE MACMILLAN

Palgrave Macmillan in the UK is an imprint of Macmillan Publishers Limited, registered in England, company number 785998, of Houndmills, Basingstoke, Hampshire RG21 6XS.

Palgrave Macmillan in the US is a division of St Martin's Press LLC,
175 Fifth Avenue, New York, NY 10010.

Palgrave Macmillan is the global academic imprint of the above companies and has companies and representatives throughout the world.

Palgrave® and Macmillan® are registered trademarks in the United States, the United Kingdom, Europe and other countries.

ISBN-13: 978–0–230–27336–8 hardback

This book is printed on paper suitable for recycling and made from fully managed and sustained forest sources. Logging, pulping and manufacturing processes are expected to conform to the environmental regulations of the country of origin.

A catalogue record for this book is available from the British Library.

Library of Congress Cataloging-in-Publication Data
Innovation policy and the limits of laissez-faire : Hong Kong's policy
in comparative perspective / edited by Douglas B. Fuller.
p. cm.
Includes bibliographical references.
ISBN 978–0–230–27336–8
1. Information technology—China—Hong Kong—Economic aspects.
2. Technological innovations—Government policy—China—Hong Kong.
3. Economic development—Technological innovations—China—Hong Kong. 4. Free enterprise—China—Hong Kong. I. Fuller, Douglas B.
HC430.T4I556 2010
338′.064095125—dc22 2010027507

10 9 8 7 6 5 4 3 2 1
19 18 17 16 15 14 13 12 11 10

Printed and bound in Great Britain by
CPI Antony Rowe, Chippenham and Eastbourne

Contents

List of Figures and Tables

Figures

Tables

Foreword

During the 1980s, thanks to its excellent geographical position, the industry of its people and the flourishing of its rules-based, free-market economy, Hong Kong won worldwide recognition as a successful "newly industrialized economy," and joined the other East Asian economies – Taiwan, Republic of Korea, Singapore – on the pantheon of the East Asian tigers. Thirty years on, the stunning economic achievements of China have eclipsed all other actors in the narrative on the East Asian economic miracle. The other three tigers – Taiwan, Republic of Korea, and Singapore – have moved on, seeking to scale new heights through greater use of technology and innovation. By contrast, with manufacturing all but extinct and services comprising 95 percent of its economy, Hong Kong has contented itself with earning a living as a regional services provider.

Many interesting questions can, and should, be asked about the road not taken in the Hong Kong story. With its neighbors and competitors all scrambling to increase the value of its economy by climbing the technology ladder, increasing substantially investment in research and development and alluring technologically advanced multinationals, Hong Kong has bowed out of the quest for technology and innovation pursued by other Asian economies. In 2007, in a reshuffle of ministerial responsibility and renaming exercise, the government signaled its lack of interest by removing the last reference to technology in the title of the relevant bureau. The government pays lip service to technology and innovation in its latest drive to create six new "industries," but its expenditure on research and development remains a measly 0.8 percent, and no effort has been made to increase expertise on technology within the administration.

The neighboring city of Shenzhen, a special economic zone of China, has ratcheted up its expenditure on research and development to 7 percent of its GDP. Singapore has spared no expenses in building biotech and other clusters. Both Taiwan and the Republic of Korea have performed better than Hong Kong in late 2009, as their technology exports provided powerful impetus to their recoveries. Hong Kong's rebound has been underpinned largely by China's spectacular recovery, the steady rise in the number of Mainland Chinese tourists, and record inflows of liquidity from Mainland China and the rest of the world.

Why has Hong Kong been such a laggard in technology and innovation, when the new Argonauts of East Asia have worked so hard to bring home the golden fleece of the Silicon Valley? Is this the inevitable outcome of Hong Kong's colonial heritage, which bequeathed such a strong trader's (and speculator's) mentality that Hong Kong people are incapable of thinking long term or breaking new ground? Or is it because the city's leadership has been so lacking in experience in investing in technology-based industries, and so hamstrung by their age-old laissez-faire, free-market philosophy, that they are incapable of telling good risks from bad, and making strategic investments? Or is it because of Hong Kong's traditional, heavy reliance on the high-yielding property sector, which has crowded out other riskier economic activities with lower returns? Or is it because Hong Kong has been so well looked after by its motherland, China, which has not failed to supply Hong Kong with tourists or liquidity whenever Hong Kong is hit with an exogenous shock, that Hong Kong has lost the motivation to reinvent itself via the technology route?

Whatever the reasons for its past apathy, the key questions for Hong Kong's belated bid for technological upgrading are: has Hong Kong got what it takes to succeed, and, second, is it too late? Hong Kong people are known for their creativity and ingenuity, and Hong Kong students have been praised for their science and math achievements. Hong Kong's universities are certainly of a sufficiently high standard to attract world-class talent such as Professor Charles Kao, the 2009 winner of the Nobel Prize in physics. But then, with its neighbors miles ahead of Hong Kong in technology investments and determination to succeed, is it too late? What does Hong Kong need to do to catch up?

To try and find answers to all these intriguing questions, Savantas Policy Institute commissioned a study on how to improve technology-based innovation in Hong Kong in 2007. The inquiries by a multidisciplinary team of international scholars were made in the course of 2007, and the findings reviewed by a team of international experts including Professor Henry Rowen of Stanford University, Professor Suzanne Berger of Massachusetts Institute of Technology, Professor Louis Pauly of University of Toronto, Professor Richard Dasher of Stanford University, and Professor Ben Martin of University of Sussex in early 2008. I am grateful to all these scholars and the editor of the final report, Dr Douglas Fuller, for their sterling efforts and valuable contributions. We hope the lessons we have learned can provide some useful pointers for the future leaders and administrators of Hong Kong, and whoever choose to go down the technology path.

Regina Ip
Legislative Councillor, Chairperson of the Board of
Governors of Savantas Policy Institute

Acknowledgments

First and foremost, the editor and other contributors would like to thank Regina Ip, chairperson of the Savantas Policy Institute. Without her recognition of the importance of innovation policy for Hong Kong's future, and her ability to muster the financial and administrative support for the large-scale, year-long research project that served as the foundation for this volume, this volume would truly never have come to fruition. Under Regina's leadership, the Savantas Policy Institute was the main institutional backer of the research and we would also like to thank the other members of Savantas' governing board for their support: Charles Chan, William Chan, Jonathan Choi, Ir Edgar KWAN Chi Ping, Alfred Kwok, Benjamin Huen Lam, Patrick Lui, Walter Mok, Professor Yun Wing Sung, and Dr Louis Shih. Alfred Kwok was particularly generous with his time and industry contacts for which we are very grateful.

We would also like to thank the other sponsors of the research behind this volume: the Innovation and Technology Commission (ITC) of Hong Kong, GroupSense (International) Ltd., Hong Kong Biotechnlogy Association, Hong Kong Computer Society, MyCard Ltd., Sun Hung Kai Properties Ltd. and Vtech Holdings Ltd.

Davey Chung of the ITC was particularly helpful as he went through the entire original draft pointing out factual errors and providing up-to-date data, and he did so without trying in any way to influence our conclusions, some of which are critical of the ITC.

The international review committee of our project (Suzanne Berger of MIT, Richard Dasher of Stanford, Ben Martin of SPRU/Sussex, Louis Pauly of University of Toronto and Henry Rowen of Stanford) provided crucial critical commentary of the individual chapters. The careful readings these five scholars gave our work was of immense help in formulating both the general themes of the volume and crafting more robust arguments in the individual chapters.

We would like to thank the many businesspeople, policymakers and academics in Hong Kong and the Pearl River Delta who gave their time, insights and other help to our research team. The following is an unavoidably incomplete list of those who have helped us so much: Central Policy Unit, Charles Mok, H. L. Yiu, Wilson Fung, Derek C. H. Louie,

Ben Cheng, Y. K. Li, Edmund T. Y. Lee, C. K. Wong, Chuck Cheng, Oliver Pun, Manuel Costa, E. Anthony Tan, S. W. Cheung, Nicholas W. Yang, Doris Kwan, Edwin Lee, Joshua Wong, Angela Mo, John Ma, Eugene Wong, Xiaogang Wu, David Zweig, Charles Sodini, Boaz Wong, Jody Wong, the Immigration Department, University Grants Council, Census and Statistics Department, and Asia Venture Capital Journal.

Even while acknowledging the tremendous amount of help we have received from Hong Kong and beyond, we would like to emphasize that we as editor and authors of this work are solely responsible for the book's conclusions, recommendations, and errors.

We would also like to thank Virginia Thorp at Palgrave for her invaluable contribution to editing this book and shepherding it to publication.

As editor, I would also like to thank Professors Chung Ming Lau and Shige Makino, who each served as chair of the Chinese University of Hong Kong's Department of Management while I was a visitor there. Their generous support provided me with the opportunity to organize and pursue the research project behind this book.

To end with an observation-cum-dedication, our team recognizes that innovation policy, like any other policy, can only succeed as far as people accept such policy as legitimate. Thus, no matter what policies and reforms we might recommend, our team of mainly outsiders cannot expect these policies to be accepted in a top-down manner. Rather, whatever policies end up working well in Hong Kong, these policies must be ones that resonate with, and gain the acceptance of, the people of Hong Kong. In recognition of this, our book is dedicated to the people of Hong Kong.

Douglas B. Fuller
King's College London

Introduction

Choosing to lose: Hong Kong's tradition of positive non-interventionism and its constraints on innovation

Douglas B. Fuller
King's College London

In *Free to Choose* (1980), Milton and Rose Friedman hailed Hong Kong's explosive growth in the immediate post-war decades as proof of the wisdom of laissez-faire. They wrote:

> In today's world big government seems pervasive. We may well ask whether there exist any contemporaneous examples of societies that rely primarily on voluntary exchange through the market to organize their economic activity and in which government is limited to our four duties (*Note*: the four duties are protection from coercion, justice, provision of public goods, and protection of those who are not responsible adults).
>
> Perhaps the best example is Hong Kong – a speck of land next to Mainland China containing less than 400 square miles with a population of roughly 4.5 million people.... The density of populations is almost unbelievable – 14 times as many people per square mile as in Japan, 185 times as many as in the United States. Yet they enjoy one of the highest standards of living in all of Asia – second only to Japan and perhaps Singapore.

Yet almost three decades after the Friedmans praised the obvious positive effects of Hong Kong's laissez-faire policies (referred to in Hong Kong as the policy of positive non-intervention), the lackluster performance of Hong Kong's knowledge economy over the intervening period calls these very same policies into question. This book explores the impact of Hong Kong's laissez-faire tradition on the efforts to build a knowledge-intensive economy to replace Hong Kong's industrial economy, which gradually moved to China in the wake of China's own economic reforms starting in 1978. In answering this question, this book does not just address Hong Kong's own particular issues but offers insight into the wider process of transforming industrial economies into knowledge economies.

In order to provide a wider scope to address the broader global issue of creating knowledge-intensive economies, the chapters in this volume use several implicit and explicit comparative lenses. The first lens uses Hong Kong's traditional peer group, the other East Asian newly industrialized economies (Korea, Singapore, and Taiwan). The second compares Hong Kong with the advanced economies of the Triad (North America, Japan, and Europe) since Hong Kong's per capita wealth is basically equivalent to that of many of these economies. The third observes Hong Kong's industries against the trends within their respective global industries.

The main finding of the book is that Hong Kong's laissez-faire tradition has crippled attempts to transform Hong Kong into a more knowledge-intensive economy. Hong Kong's lesson is one with wide applicability. Those economies not blessed with either large businesses equipped with developed innovation capabilities or well-developed institutions for fostering the creation, diffusion and commercialization of knowledge would be well advised not to follow Hong Kong's laissez-faire path. For it is precisely the lack of these attributes that have presented innovation bottlenecks that Hong Kong has been unable to break through while continuing to rely on its traditional policy of positive non-intervention. Many emerging economies face similar innovation bottlenecks, but even some of the advanced economies, especially those in recent economic decline, may face very similar constraints to Hong Kong's and may benefit from the lessons of its negative example.

There are additional themes that run through this book that apply to circumstances beyond Hong Kong's. As the "dragon head" of the Pearl River Delta, one of the most dynamic economic regions in the world, Hong Kong is very much aware of its role as regional player. Beyond the Pearl River Delta, Hong Kong is intensely involved as a major actor in two other economic regions, Greater China and the wider East Asian regional economy. At the same time, Hong Kong is a global player with its own multinationals and with an economy tightly integrated with the international economy. Hong Kong's experience of trying to be an effective regional economic actor at several levels, and a global one, gives us insight into the supposed tensions between regionalization and globalization. Simply put, are regionalism and globalism as economic strategies diametrically opposed or can some economies manage to discover some beneficial balance between the two? Many economies are grappling with this very issue because it is at the core of economic policy for a range of economies from members of formal free trade zones and customs unions, such as the EU, ASEAN, Mercorsur, and NAFTA, to ones engaged in more informal regional networks.

Hong Kong also has had to confront the effects of extensive economic integration with the pool of cheap labor, both skilled and unskilled, contained in Mainland China's economy. In dealing with this issue, Hong Kong is simply a microcosm of the wealthy economies as a whole. The advanced economies all are becoming more integrated with less wealthy emerging economies around the world from China to Eastern Europe. This integration is really what many pundits mean when they speak of globalization, so Hong Kong provides us with a test case for adjustment under globalization. One might even argue that Hong Kong provides a test case of adjustment in the face of a very advanced stage of globalization, given Hong Kong's openness and its geographic closeness to the largest, most populous emerging economy. Some, such as Richard Freeman (2004), suggest this integration may have dire consequences for wages and, consequently, for standards of living in the advanced economies. Others hold out hope for a variety of coping mechanisms that may alleviate many of the social costs of economic integration (Berger, 2005). Examining how Hong Kong has coped with integration with Mainland China adds to our understanding of the scope of possibilities for successful adjustment to globalization on the part of advanced economies.

This book does not just rely on negative examples drawn from Hong Kong's failures. The wider comparative perspectives also reveal what is working in other parts of the world in contrast to Hong Kong's own failures in building an innovation economy. Using examples ranging from Finland to Taiwan, the book provides many relevant examples of national and subnational institutions that have effectively promoted innovation in their respective economies. The volume also investigates the operation of certain broad macro-level institutions, such as university–industry linkages, in Hong Kong and through this process brings out what might approximate as global best practice by drawing on comparisons with institutions abroad. Despite its overall lack of success, Hong Kong itself also has some success stories that stand up well in international comparison. The most important lessons offered here for innovation policies are not one sided, with exhortations to learn from

Hong Kong or learn from abroad. Instead, this book's rich mix of comparative data offers deeper lessons into which policies and institutions to promote innovation may work across different national and regional contexts.

The structure of the book is as follows. The first seven chapters cover economy-wide institutions of innovation. The last five chapters examine Hong Kong's performance in five different knowledge-intensive sectors in their wider global industrial context.

The first several chapters look at some of the better performing institutions in Hong Kong. The first chapter by David Mowery compares Hong Kong's university–industry practices with successes and failures from other advanced economies, particularly the US. Poh Kam Wong's chapter examines Hong Kong's intellectual property (IP) rights regime and finds that it is quite strong, although Hong Kong has failed to nurture enough IP-related services.

In the area of skills training, two chapters debate about the performance of Hong Kong. Hart and Tian's chapter musters compelling evidence that Hong Kong has done a good job of educating its population. Wadhwa's chapter counters that intra-firm skills training in Hong Kong is very weak, even when compared with some emerging economies, such as India.

Chapters 5, 6 and 7 make more negative assessments of other critical innovation institutions in Hong Kong. Shih and Chen in Chapter 5 investigate Hong Kong's public R&D programs and find them lacking in long-term funding for research and burdened by cumbersome bureaucratic procedures. The fundamental issue is that Hong Kong insists on realizing positive, short-term returns on many of these public investments and this insistence prevents these public investments from realizing their full potential for spurring innovation in Hong Kong. Shih and Chen make an explicit telling comparison between Hong Kong's Applied Science and Technology Research Institute (ASTRI) and its original model, Taiwan's Industrial Technology Research Institute (ITRI). On virtually every aspect of operations and policies, they show ASTRI to be lagging far behind ITRI. The result is that Taiwan has a booming technology sector and Hong Kong has a tiny one.

Chapter 6 solves the puzzle of why Hong Kong has a vibrant financial sector and yet is bereft of venture capital investment directed towards knowledge-intensive start-ups. Au and White in Chapter 6 delineate the cultural and institutional barriers to such venture capital investments and utilize the historical experiences of the US and the UK to suggest reforms to spur investment in technology-intensive start-ups.

In Chapter 7, Segal addresses Hong Kong's policies towards and positioning within its various regional networks. He uncovers a lack of organizational coherence, with too many organizations involved (or none at all) and little coordination between them.

The last five chapters present case studies of Hong Kong's performance across a number of knowledge-intensive sectors. Chapter 8 looks for any potential to grow knowledge-intensive activities in what is left of Hong Kong's traditional manufacturing industries. Thun, in this chapter, persuasively argues that Hong Kong has the opportunity to become a key provider of design and intermediate goods for Chinese industries, particularly the burgeoning auto sector. In contrast, Hong Kong's firms have found it very hard to compete with the large, multinational suppliers to global auto brands. Thus, China presents a chance for Hong Kong to build global first-tier suppliers to compete with those from the Triad economies.

Chapters 9 and 10 look at one of the most promising knowledge-intensive sectors, biotechnology, and one of the largest, integrated circuits (ICs), respectively. Joe Wong in Chapter 9 argues that biotechnology as a sector has not lived up to its expectations, even in countries at the heart of biotechnological innovation, such as the United States. He then evaluates Hong Kong's progress under these globally disappointing circumstances. In contrast to biotechnology, Fuller in Chapter 10 analyzes a global innovation dispersion success story where Hong Kong historically has done well. Unfortunately, Hong Kong has fallen badly behind East Asian peers and other emerging economies in ICs from its previous promising position. This sad story just underscores the need for more proactive government policy, especially in the regional context where Hong Kong's peers have adopted just such policies.

Chapter 11 looks at creative industries where Hong Kong has traditionally been strong due to its globally renowned film industry. Tschang, in this chapter, explores another segment of the creative industries, video games, to see if Hong Kong can be successful in creative industries generally. He finds that strong institutions for supporting innovation exist, but argues that they need to be reconceptualized and reorganized in order to capture potential opportunities in these two creative industries.

The final chapter, Chapter 12, looks at the potentially burgeoning environmental technology sector. Given Mainland China's growing concerns over environmental problems, Hong Kong has real prospects for becoming a leading innovative in environmental technology if it can improve its innovative institutions and enhance cooperation with the Mainland at the same time that it fosters even stronger links with environmental technology leaders in the Triad economies. Admittedly, accomplishing this set of reforms is a tall order, but the social and economic benefits from realizing the goal of being an environmental technology platform provider are too great to ignore.

The approach of this book largely focuses on the informal and formal institutions of innovation in Hong Kong and falls in line with major approaches within both development and innovation studies that emphasize the importance of the relevant national or subnational regional institutions.[1] Works within these traditions tend to emphasize the distinctness and variety of the various national or regional capitalisms. However, this volume also addresses the other major alternative conception of the world's economy as a single system of capitalist development with the various national and regional economies playing different roles within this system. A number of chapters consider Hong Kong's place in the global innovation system that is shaped by webs of co-ethnic transnational technology communities (Saxenian, 2006) and global production networks (Ernst and Kim, 2002). In particular, the chapters on human capital development, venture capital, and individual industry sectors discuss Hong Kong's current involvement in these networks and offer suggestions on how Hong Kong can enhance its knowledge economy through new approaches to these networks. For example, Chapter 6 and several of the industry chapters advocate utilizing Hong Kong's own extensive, although largely latent, transnational network of technology entrepreneurs to spur venture capital and technology entrepreneurship in Hong Kong.

Note

1. Obviously, Hong Kong as a Special Administrative Region of the People's Republic China is a somewhat special case where the relevant state institutions are quite separate from the national institutions shaping Mainland China's innovation system and, arguably, little influenced by them. Thus, one could speak of Hong Kong's institutions as Hong Kong's national institutions rather than as subnational regional ones.

References

Berger, S. (2005) *How We Compete*. New York: Doubleday.

Ernst, D. and Kim, L. (2002) "Global Production Networks, Knowledge Diffusion, and Local Capability Formation." *Research Policy*, 31, 1417–1429.

Freeman, R. (2004) "Doubling the Global Workforce," Presentation at the Center for Economic Performance, London School of Economics.
Friedman, M. and Friedman, R. (1980) *Free to Choose*. New York: Harcourt Brace Jovanovich.
Saxenian, A. (2006) *The New Argonauts: Regional Advantage in a Global Economy*. Cambridge, MA: Harvard University.

1

University–Industry Collaboration and Technology Transfer in Hong Kong and Knowledge-based Economic Growth

David C. Mowery
University of California, Berkeley

1. Introduction

The twenty-first century is the century of knowledge-based international economic competition. More than ever, the prosperity of nations depends on the ability of public and private institutions, policies, managers, and workers to mobilize and exploit knowledge-intensive capabilities and assets. Although natural resources play an important role in economic competitiveness, the ability of even resource-rich economies to raise and sustain their citizens' living standards depends on their ability to exploit the created resources of knowledge and human capital.

Governments in both the industrial and industrializing economies have sought to address the competitiveness challenge by strengthening policies and institutions that support innovation. One such institution is the research university, and governments have launched a number of policy initiatives to enhance the contribution of national university systems to industrial innovation. Hong Kong is no exception, and the government of the Hong Kong Special Administrative Region (HKSAR) has expanded its financial support for research and "technology transfer" activities at Hong Kong universities considerably since 1997.

Four of Hong Kong's universities, Hong Kong University of Science and Technology (HKUST), the University of Hong Kong (HKU) and Chinese University of Hong Kong (CUHK) and the City University of Hong Kong, have consistently ranked among the better universities in Asia. The 2007 Jiao Tong University "Academic Ranking of World

Universities" (http://www.arwu.org/ARWU2007.jsp; accessed 2/12/08) ranked all four universities as tied for 25th place in its 2007 "Asia/Pacific" rankings, tied with Peking University and slightly behind Tsinghua (ranked 19th). The four universities were tied in the "world rankings" of the top 500 universities at roughly 200th place, behind the Tsinghua ranking of 150th and equal to Peking University's ranking. Singapore's National University was ranked as the 10th best Asia/Pacific university in the 2007 rankings, above all of the Hong Kong universities, and was tied with numerous other universities in the global rankings at roughly 100 out of 500.

Hong Kong's universities do not appear in any of the Jiao Tong study's field-specific rankings (spanning the physical sciences, the life sciences, social sciences, clinical medicine, and engineering) in areas other than engineering and computer science, where HKUST was ranked at 37, CUHK at 77, and City University of Hong Kong tied with Tsinghua at 51 out of 100 (National University of Singapore ranked 32nd in this tabulation). Hong Kong's universities have roughly maintained these positions in the Jiao Tong rankings since the rankings first appeared in 2003. The Hong Kong universities thus are leading regional universities within Asia, but are in the "second tier" of global rankings. Their status as regional but not global leaders, as well as the rough parity between the leading Hong Kong universities and the leading northern Mainland universities, raise some significant challenges for Hong Kong's universities.

In their analysis of Hong Kong's economic prospects, Berger and Lester found that a "common complaint was that the universities hold themselves aloof from the industrial sector and do not tailor their teaching and research activities closely enough to its needs" (1997, p. 66). The authors noted that a number of public initiatives had been launched by the Hong Kong government since the early 1990s to expand financial support for Hong Kong university research and strengthen university linkages with regional industry. These initiatives have expanded in number and scale since the publication of their study.

The HKSAR government has increased financial support for research collaborations involving university and industrial researchers, Hong Kong universities have expanded their efforts to support technology transfer to Hong Kong industry through patenting and licensing of faculty inventions, and universities have provided support for the formation of "spin-off" firms to commercialize faculty inventions. These initiatives appear to have produced positive overall results, although realization of their full effects will take time.

Although current policy initiatives have had positive effects, their focus on support for applied research and faculty economic entrepreneurship should be complemented by steps to strengthen the fundamental research capabilities of Hong Kong's universities. Such steps could enable faculty to become "research entrepreneurs," pursuing opportunities for support of basic research. Existing government policies also have been developed and implemented with little or no high-level strategic vision guiding their structure, goals, and evaluation. It is important for HKSAR policymakers and industrial managers alike to develop a more realistic appreciation of the nature and scope of economic benefits flowing from research universities, particularly in a small geographic and economic region such as Hong Kong.

Any "university-focused" policy to support innovation within Hong Kong should be part of a broader strategic vision for knowledge-based growth in Hong Kong and the Pearl River Delta that includes a range of policies extending beyond the region's universities. Among other things, this strategy should be coordinated with the rapidly evolving innovation policies of the People's Republic of China (PRC), which involve far larger sums of public funds and are linked to vastly greater flows of industrial investment.

2. The local environment for university–industry collaboration: R&D investment and innovation in Hong Kong industry

Understanding current policies affecting university–industry research linkages in Hong Kong requires some discussion of the economic environment within the HKSAR that has influenced the design and effects of policies toward innovation. Interviews with Hong Kong university faculty, administrators, and policymakers revealed widespread agreement that low R&D investment by Hong Kong firms provides a weak foundation for university–industry collaboration and technology transfer. Low levels of R&D investment mean that Hong Kong firms have limited "absorptive capacity" (Cohen and Levinthal, 1990) to exploit university research advances through licensing, faculty consulting, or hiring of graduates of Hong Kong universities. Hong Kong university faculty argued that the limited innovative capacities of Hong Kong's industry weakened the employment prospects within manufacturing for undergraduate and graduate degreeholders in engineering and scientific fields.[1]

Is the gloomy assessment of innovation and R&D in Hong Kong industry borne out by data on HKSAR R&D investment trends? Investment in R&D in Hong Kong appears to have grown since the mid-1990s, in part as a result of changes in public policy but also as a result of expanded industrial investment in R&D. R&D investment grew from 0.43 percent of Hong Kong GDP in 1998 to 0.8 percent in 2006, according to the Hong Kong Census and Statistics Department.[2] In spite of this growth in the share of its GDP devoted to R&D investment, Hong Kong lags well behind the PRC, as well as such neighboring economies as Taiwan or Singapore, in its R&D/GDP ratio.[3]

Although overall R&D investment in Hong Kong is comparatively low relative to the size of the region's economy, industry-funded R&D investment has grown significantly since 2002. The industry-financed share of total HKSAR R&D investment expanded from 35 percent in 2002 to 53 percent in 2006, while government-funded R&D investment declined from 63 percent to 43 percent of total R&D investment during the same period (*Statistics on Research and Development of Hong Kong, 2002 to 2006*, p. FA5).[4] As of 2006, Hong Kong's business sector performed 53 percent of total HKSAR R&D, an increase in its performance share from 33 percent in 2002, while higher education's share of overall R&D performance declined from 64 percent to 45 percent.

The Census and Statistics Department's statistical summary of Hong Kong R&D investment data reports that only 6 percent of industry-financed R&D investment in 2006 was funded by Hong Kong manufacturing firms, a decline from 19 percent in 2002, while the share of industry-financed R&D investment from firms in "Wholesale, retail and import and export trades, restaurants and hotels" grew from 37 percent to 58 percent during 2002–2006. "Import and export trades" includes a number of Hong Kong-based firms that operate production facilities on the Mainland,[5] meaning that some portion of the 58 percent of industry-financed R&D attributed to this sector is in fact linked to Mainland manufacturing industries that are managed by Hong Kong firms.

Although the share of industry-financed R&D associated with manufacturing based in Hong Kong may have declined during 2002–2006, the rising share of regional R&D financed by industry means that R&D performed in Hong Kong that is associated with financial and business services, or linked with the operation of Mainland manufacturing owned by Hong Kong-based firms, almost certainly has grown. The apparent links between PRC-based production facilities and manufacturing-related R&D activities that are based in Hong Kong underscore the significance of the relationship between Hong Kong and the Pearl River Delta region, the site of much of the manufacturing capacity managed by Hong Kong-based firms.

Some indicators also point to increased R&D collaboration between Hong Kong firms and universities, particularly Hong Kong universities, although the reported levels of collaboration display considerable fluctuation. According to the *2006 Annual Survey of Innovation Activities* compiled by the HKSAR Census and Statistics Department (2007, Table 2.11), almost 13 percent of the Hong Kong firms performing R&D in 2006 also supported "cooperative arrangements" in R&D with universities, and the vast majority (80 percent) of these "arrangements" involved Hong Kong universities (almost 17 percent involved PRC universities). The 2007 *Survey* (Hong Kong Census and Statistics Department, 2008, Table 2.11), however, revealed a sharp drop in the share of R&D-performing firms in Hong Kong that maintained such cooperative arrangements with universities to 6 percent. Hong Kong universities' share of such cooperative arrangements also dropped, to approximately 41 percent, while PRC universities' share grew to slightly more than 57 percent. The sharp swings in these survey findings suggest that these data must be interpreted with considerable caution, and it is possible that the 2007 results were affected by the early stages of the recession that began late that year. Nonetheless, the reported level of collaboration with Hong Kong universities for 2007 is consistent with the 6 percent of Hong Kong establishments reporting such collaborative arrangements with universities in the 2005 survey (see Sharif and Baark, 2006). The results of the 2007 survey also suggest increased R&D collaboration between Hong Kong-based firms and PRC universities.

Another problem in interpreting these indicators of "collaboration" is the lack of information on the size or content of individual "cooperative arrangements." Nevertheless, it is interesting that the 6–13 percent of R&D-performing establishments in Hong Kong that pursued some sort of "cooperative arrangement" with a local or PRC university in 2005–2007 is roughly comparable to the shares reported in the EU "Community Innovation Survey" (on which the Hong Kong survey is based) during 2002–2004 for such nations as Germany (8.5 percent), France (10.1 percent), or the United Kingdom (10 percent).[6] The extent of collaboration, based on this crude indicator, between Hong Kong industry and Hong Kong universities thus is not significantly below the levels observed in a number of large European economies.[7]

In summary, R&D investment from all sources within the HKSAR remains low relative to the region's GDP, especially by comparison with

R&D investment in the PRC and other regional economies. Nonetheless, HKSAR R&D investment has grown significantly during the past decade. Moreover, industry-funded R&D investment within Hong Kong has grown even more rapidly, suggesting that the "absorptive capacity" of Hong Kong firms may be improving somewhat.

3. Evolution of HKSAR policy toward university–industry collaboration

As Baark (2006) points out, until the 1990s Hong Kong's university system was small and focused primarily on education rather than research. In 1989, the British colonial governor announced plans for significant expansion in the Hong Kong university system, including the 1991 foundation of HKUST, and expansion was given additional impetus with the 2000 plan announced by Chief Executive Tung to ensure that 60 percent of the relevant age group of HKSAR citizens pursued post-secondary education of all types (including vocational training and non-degree studies). In 2007, the HKSAR government announced plans to shift the university curriculum to a 4-year undergraduate program by 2011–2012, including an expansion in the total number of faculty in Hong Kong universities by roughly 1000. This represents an increase of almost 20 percent in the 2007 full-time equivalent (FTE) faculty population of roughly 5400 in the eight publicly supported Hong Kong universities (HKUST, HKU, CUHK, City University of Hong Kong, Hong Kong Institute of Education, Polytechnic University of Hong Kong, Hong Kong Baptist University, and Lingnan University). Public funding accounts for the majority of the operating budgets of these eight universities.

A combination of low taxes and high salaries means after-tax compensation for Hong Kong university faculty historically has been generous by international standards. Several interviewees, including former Hong Kong university faculty, noted that the structure of faculty compensation provided weak incentives for Hong Kong academics to seek external research funding or to pursue opportunities for research collaboration with industry. Hong Kong faculty are paid on a 12-month basis, rather than the 9-month salary typical of faculty at US research universities, and the most important Hong Kong government research funding programs discourage the use of grant funds for teaching relief by faculty outside of the humanities and social sciences. The limited incentives for Hong Kong faculty in engineering, the physical sciences, and biomedical research to seek "summer salary" or "teaching relief" funding may weaken both fundamental research within Hong Kong universities and industry–university research links.

3.1. Public funding of Hong Kong university research

During the 1990s and early twenty-first century, Hong Kong universities expanded their research activities considerably. R&D performed in the university sector increased from 0.25 percent of HKSAR GDP in 1995 to 0.35 percent in 2006/2007.[8] Much of this growth reflected increased public funding for university research, which was an important component of the shifts in HKSAR policy toward innovation after the 1997 Asian financial crisis and political independence.

Public funding for Hong Kong university research comes from several sources. The University Grants Committee (UGC) provides block grants to the eight major public Hong Kong universities to support research staff and students. These grants are not awarded on a competitive basis through a peer-reviewed selection process. During 2006–2007, UGC block grants accounted for almost 72 percent of total research expenditures, more than HK$3870 million, at these eight universities. The Research Grants Council (RGC), established in 1992 within the UGC, oversees several programs for public funding of university research that accounted for 11 percent of these universities' research expenditures during 2006–2007. The General Research Fund (GRF) program, formerly known as the Competitive Earmarked Research Grant (CERG) program, awards grants of up to HK$1 million on a peer-reviewed, competitive basis.[9] This source of funding accounted for slightly more than HK$612 million in 2007–2008, representing 80 percent of RGC research funding.[10]

As Figure 1.1 shows, the average size of RGC grants has remained well below HK$1 million throughout the program's history, a relatively small budget for state-of-the art research in engineering or physical and biomedical sciences. As Figure 1.2 shows, the average size of GRF grants in fields such as the biomedical sciences (approximately US$140,000) is less than one-half of the average size of grants from the US National Institutes of Health. Although modest in size, these grants typically are not used to support graduate students (who are supported by universities' UGC block grants), and faculty in fields other than the humanities, social sciences, and business are discouraged from using funding from this source to "buy out" their teaching time.[11]

A second public source of research funding that specifically targets university–industry collaboration is the University Industry

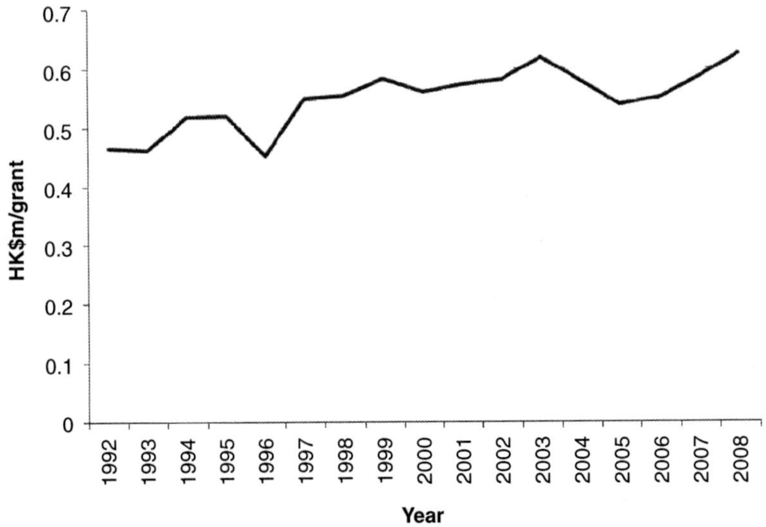

Figure 1.1 Average RGC GRF award (HK$m), 1992–2008
Source: Research Grants Council of Hong Kong, 2008.

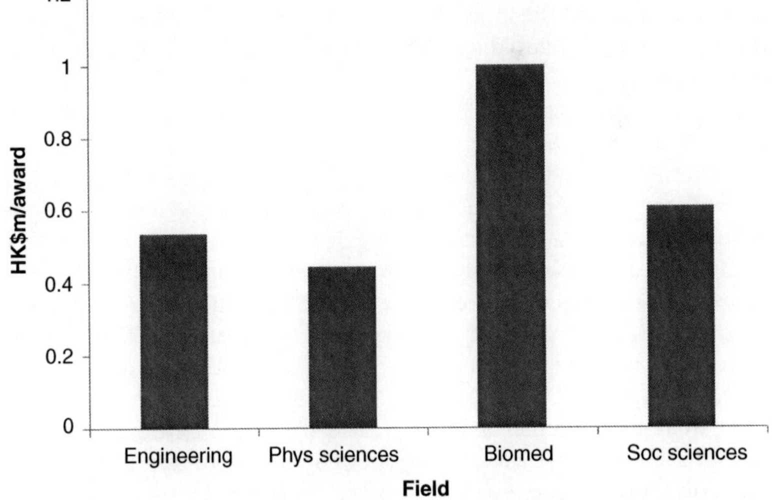

Figure 1.2 Average RGC GRF award, 2007/2008 by field
Source: Research Grants Council of Hong Kong, 2008.

Collaboration Program (UICP), established in 1999 within the Innovation and Technology Fund (ITF) which is overseen by the Innovation and Technology Commission (part of the Commerce and Economic Development Bureau, the ITC was created in 2000 as a result of the 1998 and 1999 reports of the Commission on Innovation and Technology). The UICP provides grants for research projects for which at least one-half of the total funding is provided by industry.[12] Like the GRF, however, UICP grants discourage use of funds for teaching relief,[13] and the average size of these grants, while larger than GRF grants, is modest. As of the end of 2008, 178 UICP projects had been funded, with an average contribution of ITF funds to each award of HK$1.1 million. Assuming that the ITF contribution was matched by industry funding, the average size of UICP projects is roughly HK$2.2 million, or slightly more than US$300,000.

Funding proposals to the UICP undergo separate reviews of their technical and business merits. Any intellectual property resulting from a UICP project for which industry provides 50 percent of total funding is assigned to the industrial participant. The UICP, with total expenditures of roughly HK$196 million as of the end of 2008, accounts for a small share (5.2 percent) of the overall ITF program, which has spent almost HK$3.8 billion of the HK$5 billion allocated to the program in 1999. The majority of ITF spending thus far, more than HK$3 billion, has supported the Innovation and Technology Support Programme (ITSP), which includes the "R&D Centres" (see below for additional discussion).[14]

Non-governmental sources of funding for Hong Kong university research are less significant than public funds. The Universities Grant Committee reported that during 2006/2007, 11 percent of the eight major Hong Kong universities' research funding was derived from non-governmental Hong Kong sources, much of which is likely to be industry funding.[15] But some Hong Kong universities report that industry accounts for a larger share of their research funding. Senior administrators at HKUST stated that industry funding of its research amounted to 15 percent of the institution's total research budget, a share that is well above those at UC Berkeley or Stanford, and nearly as high as the share of industry-funded research at Massachusetts Institute of Technology (MIT). And the CUHK website reported that more than 20 percent of that university's research was supported by industry during 2006/2007.[16] These high reported shares of industry-supported research are surprising, given the critical portrayals in interviews of Hong Kong industry's underinvestment in R&D and weak links with university research.

Some observers have suggested that these reported shares include ITF-funded research grants, which may mean that a portion of this reported "industry funding" of academic research in fact comes from public sources. But even if one-half of the reported "industry funding" reflects ITF funding, the reported shares of industry-funded research at these two Hong Kong universities compare favorably with the reported industry-funded shares of research budgets at Stanford and UC Berkeley. A more detailed and disaggregated analysis of data on industry funding of Hong Kong university research is needed to resolve this question. Nevertheless, these data on the shares of research at two leading Hong Kong universities that are supported by Hong Kong industry call into question some of the more critical portrayals of the commitment of Hong Kong firms to innovation and research collaboration with Hong Kong universities.

3.2. Non-university ITC programs

In the opinion of several Hong Kong university faculty and administrative interviewees, at least some of the objectives of the UICP have been undercut by the ITC's establishment of additional research initiatives that are not focused primarily on supporting academic research. In response to a 2003 assessment of the UICP that concluded that only one-half of the projects funded by the program yielded results deemed useful by industrial project partners, the ITC established a set of "targeted" R&D Centers with funding from the ITF that focused on the following specific areas:

Automotive parts and accessory systems
Logistics and supply chain management-enabling technologies
Textiles and clothing
Nanotechnology and advanced materials
Information and communications technologies
Chinese medicine.

These R&D Centers can support university research (subject to the industry matching requirement) and conduct R&D within their own research facilities.

Yet another initiative, established in 2001, is the Applied Science & Technology Research Institute (ASTRI), a government corporation that is independent of the ITC. The ASTRI was established to support "downstream" R&D that could complement academic research on industrial technologies, facilitating the transfer and commercial application of academic research results. ASTRI, which now houses the ITC R&D center in information and communications technologies, works with Hong Kong industry on applied R&D to develop commercially applicable prototypes of technological advances. As Baark (2006) notes, the ASTRI was modeled on Taiwan's Industrial Technology Research Institute (ITRI), which has been credited with aiding the growth of Taiwan's industrial capabilities in semiconductors, IT, and advanced materials.

3.3. Assessment

The establishment of the Research Grants Council (RGC) in 1992 represented the first large-scale public program in Hong Kong to support competitive grants for academic research. Nevertheless, the majority of the funding for university research provided by the HKSAR government is allocated on a non-competitive, "block grant" basis by the UGC. Grants awarded by the RGC on a competitive, peer-reviewed basis account for a small share, almost certainly less than 10 percent, of total Hong Kong university research funding from all sources. Moreover, the average size of these grants is small by comparison with those awarded by industrial-economy governments, and use of these grants by engineering and science faculty to release time from teaching is discouraged. In order to support the development of a globally competitive knowledge-based economy, more public financial support for competitively awarded, long-term research grants of greater average size is essential. Enabling faculty to use grant funds from the RGC to free up research time might enhance the incentives for faculty to seek research funding from this source and strengthen long-term research capabilities within Hong Kong universities.

None of the Hong Kong government programs providing financial support for university–industry research collaboration and innovation within Hong Kong industry is even a decade old, and more time is needed to evaluate their effects. Increased government funding for academic research linked to industry during the past 10 years has been associated with growth in industry-financed R&D investment, but it is impossible to determine whether public R&D funding has "catalyzed" greater industry R&D investment. Recent growth in industry-financed R&D appears to be concentrated in industrial sectors that have received less public financial support, although the available data do not enable a more detailed assessment.

A number of university researchers characterized the UICP grants as too small to support ambitious research projects with potentially high long-term payoffs. In addition, the administrative and paperwork "overhead" associated with these grants, which are disbursed in 6-month installments contingent on detailed progress reports, was criticized by many researchers as burdensome and a deterrent to seeking research funding from the UICP. University faculty also criticized the slow pace with which the ITC had expended its budget, pointing out that nearly one-quarter of the ITF's original funding allocation remains unspent after nearly 10 years of operations. The prohibition on using UICP grant funds for marketing and design activities also was criticized by several recipients of these grants.

Researchers criticized the review process for UICP funding proposals as prone to bias and conflicts of interest. In their view, the review of the economic and business aspects of UICP funding proposals frequently involved Hong Kong industry personnel who, in some cases, had private incentives to deny funding to a project that might benefit a competitor, and in other cases might imitate the technical development described in funding proposals. As a result, a number of faculty noted that they no longer sought funding from the ITF's UICP program, and many expressed interest in obtaining funding from PRC firms or even public sources, recognizing that these funds could not be disbursed through their Hong Kong universities.

With the exception of the RGC, the research funding initiatives undertaken by the HKSAR government since the late 1990s focus primarily on supporting industry–university collaborative research, much of which necessarily will focus on applied research. There are few opportunities for Hong Kong university faculty to obtain large grants for long-term, fundamental research, which is less likely to receive support from industry. But the long-term strength of Hong Kong's universities as research institutions, as well as their attractiveness to first-rate academic researchers, depends on their capabilities in fundamental, as well as applied, research. As the research capabilities of PRC universities, particularly in southern China, improve, firms in the Pearl River Delta region and elsewhere that seek opportunities for collaboration with university researchers in applied fields may collaborate with local Mainland universities. At the same time, global firms (including some from the PRC, such as Huawei) seeking academic research expertise in fundamental areas will look for universities in Europe and North America, unless Hong Kong universities can build their strengths in fundamental research. The current focus of the larger-scale research funding programs operated by the ITC on applied research thus should be complemented by expanded support for longer-term research at the frontiers of knowledge, and such support should operate through competitive awards that are subject to peer review by international experts.

Both the ASTRI and "R&D Centres" programs also were criticized by university administrators and researchers, for several reasons. The most common (and unsurprising) criticism noted that these organizations were competitors for ITF funds that might otherwise support faculty research. These programs were criticized for their ineffective links with university researchers, and HKUST faculty in particular felt that collaboration with ASTRI, a necessary precondition for the research organization to fulfill its mission as a mediator between Hong Kong industry and university research, was ineffective, reflecting conflicts over intellectual property, funding, and other issues.

Finally, faculty and administrators criticized the ITF and related initiatives as being fragmented and poorly integrated with a strategy for Hong Kong's knowledge-based growth. The lack of such a strategy is reflected in the current planning for the allocation of the large number of new faculty positions to support the shift to a 4-year undergraduate curriculum in 2011–2012.

These additional faculty positions represent an increase of nearly 20 percent in the overall number of current Hong Kong university faculty, and the allocation of these faculty positions among universities, research fields, and disciplines presents an opportunity to strengthen the academic research capabilities of Hong Kong's universities. Current proposals for allocating these slots, however, appear to emphasize reliance on projections of undergraduate enrollments by subject area, along with a strong preference for "fair shares" among the eight universities. Allowing enrollments and equity among Hong Kong universities to be the primary criterion for allocating all of these faculty positions, however, may impede the exploitation of an opportunity to enhance the research capabilities of Hong Kong universities in selected areas. Setting aside 10–25 percent of these positions to be allocated among Hong Kong universities on a competitive basis in response to proposals from campuses (or multicampus proposals) for creating or strengthening areas of research excellence seems well worth exploring. The lack of a more strategic approach to the allocation of this large investment of public resources in Hong Kong universities underscores the broader failures of planning and vision within the agencies overseeing the universities and the HKSAR area's economic development.

4. Hong Kong university technology transfer programs

At least five of the eight UGC-supported Hong Kong universities (HKUST, Polytechnic University of Hong Kong, CUHK, HKU, and City University of Hong Kong) began programs in patenting and licensing faculty inventions, along with various complementary initiatives to support faculty spin-off firms, during the 1980s and 1990s. This section discusses the evolution of each university's program and then examines overall trends in the quantity and quality of Hong Kong university patents during the 1986–2008 period.

4.1. Evolution of Hong Kong universities' technology transfer strategies

All five of the major Hong Kong research universities active in patenting (HKUST, CUHK, HKU, City University of Hong Kong, and Polytechnic University) also operate programs to support faculty entrepreneurship and research collaboration with industry. The technology transfer strategies of these five universities differ considerably, however, and have changed during the past 20 years. For example, during the 1990s Hong Kong University established an independent technology licensing firm, Versitech, to support the formation of spin-off firms to exploit HKU intellectual property. In addition, HKU established an "incubator" to house and support faculty-founded spin-off firms, although interviewees indicated that HKU is now shrinking its incubator facility. The university founded its Office of Technology Transfer in 2006 to provide support for faculty patenting and licensing, and now places greater emphasis on licensing intellectual property to established firms, including patent "aggregators," which are firms that seek to accumulate patent portfolios to license to other firms and/or to use as a basis for patent infringement suits.

By contrast, CUHK funded its Office of Technology Transfer in 1991. Although CUHK has no formal university-supported incubator facility for supporting faculty spin-off firms, various academic departments and research organizations (such as the Center for Innovation and Technology) provide informal support for faculty entrepreneurship. In addition, CUHK held equity positions in 18 spin-off firms as of January 2008. Other CUHK interviewees, however, argued that the university has been reluctant to approve faculty applications for leave of absence to work in spin-off firms, and further claimed that the university's Office of Technology Transfer now focuses its licensing efforts on established, rather than spin-off, firms.

HKUST combines a large patenting and licensing program with support for faculty entrepreneurship, including a relatively liberal leave of absence policy for faculty, a university incubator for spin-off enterprises, and, in some cases, financial support for these spin-offs through the university's Research and Development Corporation (RDC). At the same time, however, HKUST has licensed patents to a number of established US and European firms, and also licenses its intellectual property to patent aggregators. Like many US universities that entered into direct management of patenting and licensing activities after 1980, HKUST's original patenting "strategy" was not selective, resulting in the issue of a number of low-quality patents to the university and high operating costs for the university's Office of Technology Transfer. The university now appears to have adopted a more selective approach to patenting which has reduced emphasis on licensing revenues and instead seeks to support faculty entrepreneurship and attract research support from industry.[17]

The City University of Hong Kong and Polytechnic University of Hong Kong have pursued a different mix of activities in technology transfer. Both of these universities support patenting and licensing of faculty research, as well as faculty entrepreneurship. In addition, however, both universities operate programs that manage faculty consulting and share in consulting revenues. Like the three other universities discussed above, both Polytechnic University and City University have changed the mix and emphasis of their policies during the past 15 years.

City University established an incubator in the early 1990s, which was overseen by the university's Technology Transfer Office (TTO) as part of a broad policy that relied on faculty-founded start-up firms as vehicles for licensing and commercialization of university intellectual property. City University's technology transfer operations were reorganized in 2003, shifting the incubator out of the TTO and reorienting the university's licensing policy to focus more intensively on established Hong Kong firms as licensees. Interviews suggested that this new approach was motivated by a growing concern within the university administration over potential conflicts of interest and commitment associated with faculty entrepreneurial activity, as well as the limited success of the start-ups "spun out" from City University, which, in turn, was attributed in part to the modest management skills of faculty entrepreneurs.[18] City University also operates a "Professional Services" organization that markets and manages faculty consulting, promoting faculty expertise to industry in Hong Kong and elsewhere.

The university receives 25 percent of gross consulting receipts as an administrative fee for providing this service.

The experience of Polytechnic's technology transfer programs is similar in many respects to that of City University. Like City University, Polytechnic manages and promotes faculty consulting activities (which have been especially significant in construction engineering and textiles production), charging a fee against gross faculty receipts. Polytechnic also has shifted its technology transfer strategy away from its early emphasis on faculty-founded spin-off firms, based on concerns over conflicts of interest and disappointment with the limited success of its spin-off firms. The university imposes stringent reporting and financial oversight requirements on spin-off firms in which it holds equity, which (in the view of the current TTO director) have discouraged such entrepreneurial activities, and some attention now is devoted to relaxing somewhat these requirements. The current patenting and licensing efforts of Polytechnic are focused on established firms, although the TTO has had limited success in attracting non-Hong Kong firms as licensees.

4.2. Hong Kong universities' patent activity, 1985–2008

Data on United States Patent and Trademark Office (USPTO) filings by Hong Kong universities (Figure 1.3)[19] indicate that university-assigned patents were almost nonexistent before the 1990s (one patent was applied for in 1986 by Hong Kong University), but applications grew rapidly during the 1990s, peaking during the 1996–2000 period.[20] Although it is tempting to conclude that such expansion in applications for patents by Hong Kong universities reflects increased faculty engagement in technology-transfer activities, these trends capture only growth in patents assigned to Hong Kong universities. It is possible, for example, that Hong Kong university research has long formed the basis for patent applications by firms based in Hong Kong that may have hired faculty as consultants or otherwise supported their research in exchange for assignment to these firms of any patents resulting from such research. Hong Kong university faculty thus could have been heavily engaged with industry in applied research activities that did not produce university-owned patents.

This type of faculty–industry collaboration has been significant in Western Europe. Recent work by Crespi et al. (2006) suggests that less than 25 percent of the European Patent Office (EPO) patents applied for during 1993–1997 for which university faculty in Great Britain,

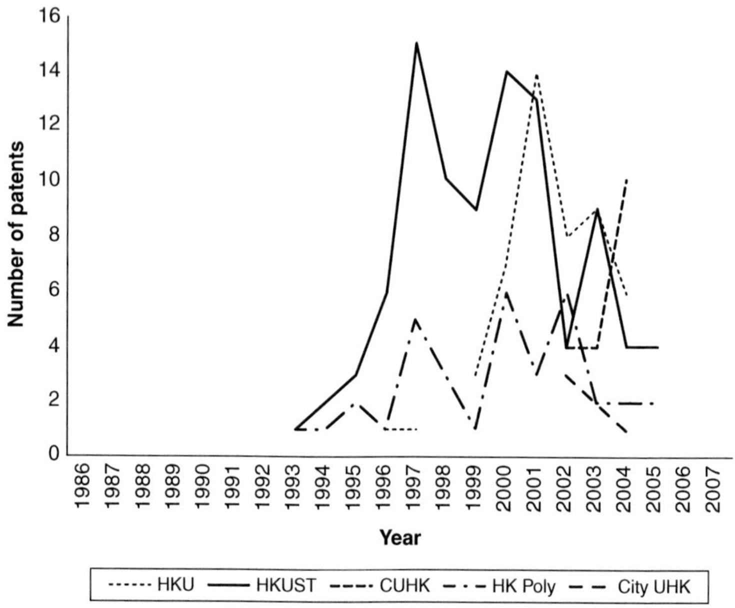

Figure 1.3 Hong Kong universities patenting, by year, 1986–2007
Source: USPTO (http://patft.uspto.gov/netahtml/PTO/search-bool.html; accessed 15 September 2008).

Netherlands, France, Germany, Italy, and Spain are listed as inventors or co-inventors were assigned to their universities, suggesting that considerable European university faculty patenting has not been captured by data on university-assigned patents. A more detailed analysis of the patents assigned to Hong Kong firms could shed light on the importance of similar faculty–industry interactions, if any, involving Hong Kong-based faculty. Certainly, the data on Hong Kong university patenting must be interpreted with great caution in light of this and other complexities, and "reforms" that seek primarily to expand Hong Kong university patenting must be scrutinized carefully.

Interpretation of the growth in overall patenting by Hong Kong universities, particularly HKUST, CUHK, and HKU, during the past 15 years also must be tempered by recognition that the technological and economic importance of individual patents varies widely. The experience of many US universities that expanded their patenting significantly after the passage of the Bayh-Dole Act suggests that practical experience is needed in order to develop a strategic approach to patenting that produces high-quality patents (see Mowery et al., 2002). With this historical experience in mind, it is worth examining the "quality" of Hong Kong universities' patents, focusing on USPTO patents applied for during 1986–2006.

The usual measure of patent quality is the average number of citations to a patent made in subsequent patent applications. Citations to patents serve as a means for an applicant to differentiate their technical advance from previously patented inventions, and therefore it is important for applicants to base citations to prior patents on a systematic search of previous patents. Patents of greater technological and economic value tend to receive more citations, on average, in subsequent patents. Figure 1.4 compares the average number of these "forward citations" received by Hong Kong universities' patents with those for USPTO patents filed by US and Singaporean universities (Nanyang Technical University and the National University of Singapore). Since patents issued earlier have a greater period of time during which they can be cited than more recent patents, the citations data cover only the first 5 years following the date of issue of a patent. This sample of patents therefore includes only

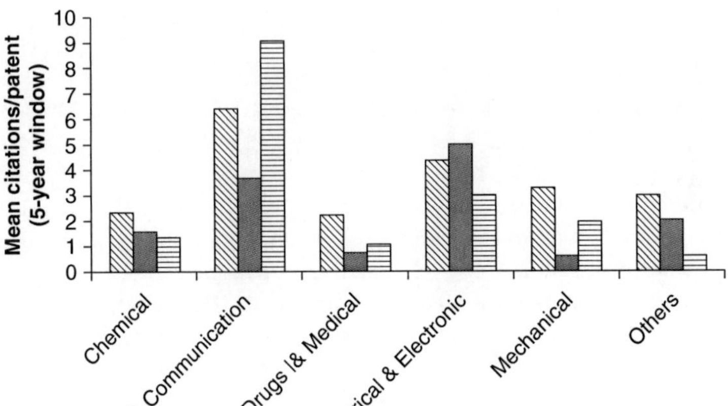

Figure 1.4 Mean forward citations/patent, 1986–2003 patents assigned to US, Singaporean, Hong Kong universities

Source: USPTO (http://patft.uspto.gov/netahtml/PTO/search-bool.html; accessed 15 September 2008) and Hall et al., 2001.

those applied for during 1987–2003 and issued during 1988–2003 (citations extend through 2008). In order to control for differences among broad technology classes in the propensity to patent and to cite previous patents, citations data also are disaggregated among five technology classes (chemicals; computers and communications technologies; drugs and medical technologies; electrical and electronics technologies; and mechanical technologies). The data include only citations to patents made by entities other than the assignees.

The data in Figure 1.4 indicate that Hong Kong universities' US patents are cited less frequently in other issued US patents than those assigned to US universities in all technology classes other than "computer and communications technologies." But Hong Kong universities' patents in this technology class are cited more heavily on average than those issued to either US or Singaporean universities. Moreover, Hong Kong universities' patents in the "drugs and medical" and "mechanical" technology classes are cited more heavily on average than US patents from the same time period issued to Singaporean universities, although the number of patents in individual comparisons is so small that these differences are not likely to be statistically significant.

Computers and communication technologies account for almost 24 percent of the patents issued to Hong Kong universities included in this analysis (Figure 1.5), a larger share than that for the US or Singaporean universities included in Figure 1.5 (7 percent and 19 percent, respectively). These results thus suggest that in one of the technological fields in which they have been most actively patenting, Hong Kong universities have patented inventions of high economic or technological value. But in the technology field that accounts for the greatest single share of Hong Kong universities' patenting, electrical and electronics technologies (which accounts for 25 percent of Hong Kong universities' US patents applied for during 1987–2003), Hong Kong universities' US patents are cited less heavily on average than those issued during the same period to US and Singaporean universities.

4.3. Hong Kong university links with PRC firms and R&D funding programs

Many Hong Kong universities and faculty have sought to develop closer links with PRC universities, firms, students, and government R&D programs during the past 15 years. These links have expanded considerably, although they are not captured in publicly available data. HKUST has established a research facility on the Mainland, CUHK has established

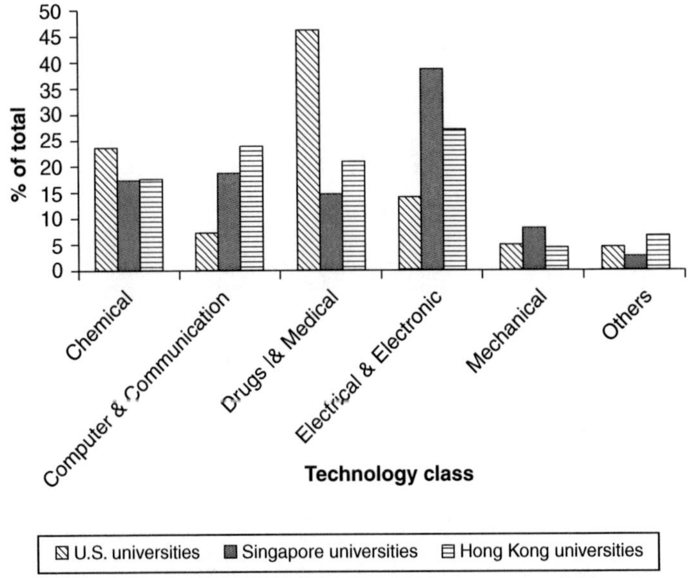

Figure 1.5 Technology class shares, 1986–2003 patents assigned to US, Singaporean, Hong Kong universities

Source: USPTO (http://patft.uspto.gov/netahtml/PTO/search-bool.html; accessed 15 September 2008) and Hall et al., 2001.

an industrial liaison office in the PRC, and other Hong Kong universities are active in teaching programs and other types of industry outreach in the Pearl River Delta region. Enrollment of PRC citizens in Hong Kong university graduate programs also has expanded, although immigration restrictions have limited growth in undergraduate enrollment of PRC nationals in Hong Kong university education.

PRC government spending on R&D has grown rapidly since 2001,[21] and interviews indicated that a number of leading Hong Kong university researchers had obtained funding from PRC government R&D programs. But these funds cannot be used to defray research expenses incurred within their Hong Kong universities. HKUST administrators expressed interest in obtaining PRC government funding for their Shenzen research institute, although no grants have yet been awarded. Finally, major PRC firms, notably Huawei, have enlisted a number of Hong Kong university faculty (particularly HKUST faculty) in collaborative research projects based in the PRC. A number of interviewees, however, noted that PRC nationals employed in the R&D operations of these Mainland firms often face obstacles to travelling to the HKSAR for consultations or visits to Hong Kong university research facilities.

The current scale of public R&D funding supported by the PRC government outstrips the R&D spending of the HKSAR government. Moreover, both public and industry investment in R&D within the PRC is likely to grow more rapidly during the next decade than will be true within the HKSAR. Entrepreneurial faculty and university administrators will continue to seek to tap these funds, drawing on the established strengths of Hong Kong's universities in research and education. It seems inevitable that this process will link Hong Kong universities more and more closely with PRC firms and R&D programs, regardless of the posture of HKSAR government policy. Indeed, the data on collaboration between Hong Kong industrial firms and universities cited earlier seemed to indicate rapid growth in the number of such collaborative R&D arrangements with PRC universities during 2006–2007. These stronger links between Hong Kong universities and PRC industry, universities, and government programs are likely to benefit Hong Kong industry, since so much of the PRC's manufacturing activity is managed or owned by Hong Kong firms that are significant sources of R&D investment within Hong Kong. But a failure to respond creatively to these growing links may limit their benefits for Hong Kong citizens and universities.

4.4. Assessment

The contrasting evolution of the "entrepreneurship" and technology-transfer policies of these five leading Hong Kong universities is striking. Three of the universities (City University, HKU, and Polytechnic University) have substantially reduced their institutional support for faculty entrepreneurship (including incubators), while HKUST has expanded its efforts in this area and CUHK's formal institutional support for faculty-founded spin-off firms has been more limited. One reason for reductions in support for faculty-founded firms is the modest success enjoyed by Hong Kong university spin-off firms, only three of which (EcoTek, Tele Eye, and Vertex) have been listed on the Hong Kong Growth Enterprise Market. The recent expansion of support for faculty entrepreneurship at both HKUST and CUHK does not yet appear to have expanded the number of spin-off firms from these universities that have been listed on the Hong Kong Growth Enterprise Market.

Hong Kong's universities have been active in patenting for less than two decades, but the university administrators overseeing these activities have adopted a realistic view of the likely pay-offs from university patenting and licensing. In interviews, Hong Kong university officials did not portray patenting and licensing as potentially important future sources of revenue. Instead, patenting and licensing were viewed as one of several ways to enhance links with industry and expand the contributions of university research to regional economic development.[22] As such, it is unreasonable to focus on the number of patents, the number of licenses, and/or the inflow of licensing royalties as indicators of the "success" or "failure" of this aspect of Hong Kong universities' efforts to transfer technologies to industrial practice.[23] At the same time, interviewees argued that few established Hong Kong firms had sufficient internal expertise to understand the potential payoff to licensing of university technologies, and/or lacked the expertise to manage inward licensing effectively.[24]

Overall, Hong Kong universities' patenting performance has been fairly effective, keeping in mind that they have pursued significant levels of patenting for no more than 10–15 years. The fact that patents in the technology category accounting for the second largest share of Hong Kong universities' patenting are more heavily cited on average than those assigned to US universities during this period is also noteworthy.

As I noted earlier, both CUHK and HKUST have negotiated licensing agreements with so-called patent "aggregators," enterprises that seek to construct portfolios of patents to license as packages to other firms. Such licensing agreements may generate significant licensing fees, but do not always result in the successful application of the technologies under license. In the United States, a number of patent aggregators have behaved as "patent trolls," seeking to develop a position that blocks other firms from pursuing innovations in a given field and litigating aggressively. The "Nine Principles to Consider in Licensing University Technology" statement issued by a group of leading US research universities in 2007 expressed ambivalence over licensing agreements with patent aggregators, noting that "universities would better serve the public interest by ensuring appropriate use of their technology by requiring their licensees to operate under a business model that encourages commercialization and does not rely primarily on threats of infringement litigation to generate revenue."[25] Licensing to aggregators may produce near-term income, but could reduce the possibilities for using patent licenses as an instrument to obtain research funding from industry, especially if potential research sponsors view aggregators holding licenses from these universities as a threat to their business operations.

Another issue that merits consideration in any discussion of Hong Kong universities' patent licensing activities is the value of interinstitutional collaboration in managing patenting and licensing activities. The annual flow of faculty inventions, patent applications, and licenses at each of the leading five Hong Kong universities is small, and the costs of staffing and managing technology transfer offices are high. Moreover, a share of these operating costs (e.g., patent prosecution costs, or the expenses of the necessary minimum staff of licensing professionals) does not vary with the volume of disclosures, patent applications, or licenses. The cost structure of technology licensing operations, along with the limited supply of expertise and the strong evidence that considerable "learning by doing" can improve the efficiency and effectiveness of these offices, all represent strong arguments for greater collaboration among Hong Kong universities regarding their technology licensing activities. Some of the activities of these licensing offices, such as working with faculty in the development of patenting and licensing strategies for a given invention, almost certainly must remain decentralized at the individual university level. But other activities, such as applying for patents or defending these patents, might well be shared among Hong Kong universities.

There also may be greater scope for informal informational exchange among administrators at these universities who are seeking to expand university–industry collaboration. Moreover, the declining institutional support for entrepreneurial activities at three of these five universities could provide an opportunity for expanded initiatives spearheaded by the Hong Kong government, by a cooperative effort managed by HKU, Polytechnic University, and City University, or by these three universities and the Hong Kong "Cyberport" (which currently operates an incubator and provides other services for start-up firms), supporting a regional incubator facility that might better address the continuing interest of faculty and industry in entrepreneurial activities.

There is little evidence of any comprehensive strategy within the HKSAR government on the management and encouragement of links among PRC programs, firms, or universities and Hong Kong universities. The modest R&D initiatives sponsored by the HKSAR government do not appear to be coordinated with the priorities and goals of the far larger PRC R&D investment programs.[26] Moreover, opportunities for expanding enrollment of PRC nationals within professional-degree and undergraduate programs within the HKSAR could be exploited more extensively. Graduate students, postdoctoral research fellows, and even master's degree students are important channels for the development of research linkages between industrial and academic researchers, and obstacles to PRC nationals pursuing these opportunities within Hong Kong universities reduce the prospects for collaborative research links with Mainland firms. Restrictions on travel by PRC R&D professionals to Hong Kong universities for collaborative research also increase the incentives for Hong Kong university faculty and administrators to shift their R&D to the Mainland.

5. Conclusions

Hong Kong's public universities have been the focus of a series of policy initiatives during the past decade that seek to enhance university–industry collaboration and technology transfer. These public initiatives have complemented and catalyzed efforts by Hong Kong universities to expand their patenting, technology licensing, and "faculty entrepreneurship" activities. In light of the fact that all these initiatives are fairly recent, it is important that policymakers and other interested parties exercise patience in the evaluation and (as appropriate) modification of these initiatives and policies. More time is needed to assess the effects of these shifts in policy and expectations on the role of Hong Kong's universities in knowledge-based growth within the region.

At the same time, however, the broader environment within which Hong Kong's universities operate also is changing as a result of the development of the PRC economy and expansion in PRC government programs to strengthen knowledge-intensive growth on the Mainland. Greater effort should be devoted to coordination of the HKSAR and PRC (as well as regional governments') strategies for economic development and knowledge-based growth in the Pearl River Delta region. In addition, the efforts of HKSAR policy to encourage faculty entrepreneurship in starting new firms and commercializing their discoveries should be complemented by expanded efforts to develop "research entrepreneurship" within Hong Kong universities, by increasing these grants' size, flexibility, and financial benefits for faculty.

Indeed, the transformation of the PRC economy and innovation system means that the focus of Hong Kong government policy on supporting university–industry collaboration in applied research needs to be re-examined. Although funding for applied research is important, the fundamental research capabilities of Hong Kong's universities also require support. Hong Kong universities currently are ranked as roughly equal to the best PRC universities (Peking University and Tshinghua University), and the quality of these Mainland institutions is likely to improve significantly as a result of increased PRC government spending on research and education. Moreover, the universities in the Pearl River Delta, which lag behind Hong Kong universities in research capabilities, may well narrow this gap in the future. Merely maintaining the position of Hong Kong's universities relative to the best PRC universities thus is likely to require improvement in Hong Kong universities' fundamental research capabilities. And support for this type of longer-term research necessarily will have to come largely from public sources. Fundamental research support is less likely to attract industry collaborators and will require grants of larger scale and longer duration.

At present, HKSAR research support for universities relies too heavily on non-competitive block grants, and the competitive, peer-reviewed grants that are supported by the RGC are too small on average to strengthen the fundamental research capabilities of Hong Kong university faculty. Moreover, the rigid policies governing RGC grants limit their attractiveness for faculty and may weaken the attractiveness of Hong Kong university positions for world-class researchers.

The attempts of Hong Kong's universities to expand their efforts in technology transfer and support for faculty entrepreneurship have enjoyed mixed success. Nonetheless, after less than 20 years' experience in patenting and technology licensing, university administrators at all five of the Hong Kong universities active in these areas emphasized the use of patents and technology licensing to achieve a broader set of goals than licensing income alone.

At the same time, the emphasis in many Hong Kong universities on faculty-founded start-up firms as a solution to the perceived weaknesses of established Hong Kong firms as research collaborators or licensees for university-developed technologies may be unrealistic. Although a few anecdotes about successful cases have received enormous attention, US universities have had mixed results with spin-off-centered technology transfer strategies (spin-off firms account for a relatively small share of overall US university technology licensees – see AUTM, 2001, 2002), reflecting the high mortality rate of new firms generally, as well as the limited managerial talents of faculty entrepreneurs. The modest supply of venture capital for new enterprises in Hong Kong means that the possibilities for transformation of the Hong Kong and regional economies through university-spawned start-up firms may in fact be more limited than is true of the United States.[27] Although the efforts of Hong Kong universities to support faculty entrepreneurship should be maintained, expectations concerning the long-term transformative effects of these strategies on the regional Hong Kong economy must be tempered, and the emphasis on "economic entrepreneurship" among Hong Kong university faculty must be balanced by policies that create great incentives and opportunities for faculty "research entrepreneurship."

Indeed, a central assumption of the "spin-off-focused" technology transfer, the belief that Hong Kong industry lacks the necessary "absorptive capacity" to collaborate with Hong Kong university researchers and/or hire technical degreeholders, is open to question. Data covering the past decade indicate considerable growth in industry-funded R&D investment within Hong Kong, and the Hong Kong Innovation Survey suggests that Hong Kong firms' self-reported levels of collaboration with local universities may be comparable to those of Western European firms. Nor is the tendency for technical graduates of Hong Kong universities to seek employment in non-manufacturing industries unique to Hong Kong. Either these data fail to measure the characteristics of Hong Kong firms that discourage their reliance on an innovation-oriented competitive strategy, in which case better data are needed, or the data are accurate and the repeated criticisms of Hong Kong industry's failings reflect other impediments to more effective university–industry interaction. In either case, a more detailed survey of incentives and impediments to research collaboration and technology transfer would be invaluable in developing a clearer analysis of university–industry collaboration within Hong Kong and the PRC.

Finally, any policy to improve the contributions of Hong Kong universities to knowledge-based growth within Hong Kong should be part of a more coherent HKSAR strategy to support innovation. The absence of such a strategic vision has hampered the integration of policies to enhance the contributions of Hong Kong universities to technology-based growth with other industrial policies of the HKSAR. The absence of a clear technology strategy for the HKSAR also impedes coordination with the rapidly evolving (and generously funded) portfolio of central and regional government policies on the Mainland.

Notes

1. Indeed, several faculty interviewees noted that many technical degreeholders currently find more lucrative employment prospects in Hong Kong's financial services industries.
2. http://www.censtatd.gov.hk/hong_kong_statistics/statistics_by_subject/index.jsp?subjectID=7&charsetID=1&displayMode=T; accessed 29 August 2008.

3. The most recent Organization of Economic Cooperation and Development (OECD) data list R&D/GDP shares for the PRC, Singapore, and Taiwan in 2004 respectively as 1.23 percent, 2.25 percent, and 2.56 percent (*OECD Science, Technology, and Industry Outlook 2006*, p. 210).
4. http://www.censtatd.gov.hk/freedownload.jsp?file=publication/feature_article/B70805FA2008XXXXB0100.pdf&title=Statistics+on+Research+and+Development+of+Hong+Kong%2c+2002+to+2006&issue=-&lang=1&c=1; accessed 29 August 2008.
5. "many establishments previously engaged in manufacturing relocated their labour-intensive manufacturing processes to the Mainland of China through sub-contract processing arrangement, leaving in Hong Kong only the higher value added activities like product design and R&D." (*Statistics on Research and Development of Hong Kong, 2002 to 2006*, p. FA8).
6. See Gulbrandsen and Nerdrum, 2009.
7. The economy with the highest incidence of such "collaborative arrangements," Finland, reported that 33 percent of R&D-performing establishments had "collaborative arrangements" with universities (Gulbrandsen and Nerdrum, 2009).
8. http://www.ugc.edu.hk/eng/ugc/publication/report/figure2007/15.htm; accessed 25 August 2008.
9. The RGC website notes that the lower threshold for funding applications is HK$150,000 in Engineering, Physical Sciences, and Biology and Medicine, and HK$100,000 for applications in Humanities, Social Sciences, and Business Studies: "There are no upper limits, but applicants/institutions should appreciate that given the considerable competition for the limited funds available, justifications for projects costing over $1 million will need to be particularly well argued and supported." (http://www.ugc.edu.hk/eng/rgc/grf/application/cergia.htm; accessed 18 August 2008)
10. http://www.ugc.edu.hk/eng/ugc/publication/report/figure2007/15.htm; accessed 25 August 2008.
11. "The RGC agrees in principle to provide, in cases where there is genuine need, funding for relief teachers so as to enable the PI to allocate sufficient time for research. Relief teachers engaged for this purpose are meant to relieve the PIs of their day-to-day teaching loads and administrative burden related to teaching work. Nevertheless, the RGC is of the view that it is the primary responsibility of the institutions to put their resources in areas where they would be best used. Hence, **such funding will be provided only exceptionally and upon detailed and sound justifications**." (emphasis in original; http://www.ugc.edu.hk/eng/doc/rgc/form/GRF2.pdf, p. 9; accessed 30 October 2008)
12. In its October 2007 "Consultation Paper," the ITF announced its intention to allow universities based outside of Hong Kong to participate in the UICP (http://www.itf.gov.hk/eng/Forms/Consultation_Paper_on_UICP.pdf; accessed 19 August 2008).
13. "Unless otherwise agreed to by ITC [Innovation and Technology Commission], the ITF will not pay any emolument to (i) the existing staff of the company and (ii) staff members who are already on the payroll of a university. This principle should apply irrespective of whether the

relevant service/work is carried out within or outside normal working hours of the person concerned." (http://www.itf.gov.hk/eng/Forms/itf-uicp-guide.pdf (pp. 26–27); accessed 30 October 2008).

14. http://www.itf.gov.hk/eng/statistics/StatTable101View.asp?StatTypeId=101&StatId=516&StatCaption=Statistics+of+Approved+Projects; accessed 27 October 2008.
15. http://www.ugc.edu.hk/eng/ugc/publication/report/figure2007/15.htm; accessed 25 August 2008.
16. http://www.cuhk.edu.hk/iso/facts/issue/2008/research_e.htm; accessed 29 September 2008.
17. As I noted earlier, HKUST administrators reported that industrial sources accounted for 15 percent of the university's research budget, more than three times the share observed at either Stanford University or UC Berkeley.
18. Interestingly, the co-founder of one of City University's more successful "spin-off" firms (Tele Eye) stated that he had not used the City University incubator in the early years of his firm's development in order to avoid the development of an organizational culture influenced by academic, rather than industrial, norms of behavior. The co-founder stated that City University TTO administrators did provide useful contacts and guidance, particularly in the firm's initial public offering (IPO). But overall, this individual stated a preference for an incubator infrastructure in Hong Kong supported by the HKSAR government rather than by universities (Interview, Cliff Chan, 10 June 2008).
19. Are US patent data the most reliable indicators of Hong Kong universities' patenting? Hong Kong universities can file for patents in the HKSAR patent office, although this is little more than a registration system, involving very limited examination of applications; they can also file for PRC patents, for European Patent Office patents, or for USPTO patents, among other alternatives. It is plausible, given relative market size and historic commercial links, that the USPTO patent trends are representative; but this is an assumption rather than a fact. It is also plausible, however, that given the lower costs of filing for patents in the "home country" of a given university, the USPTO patents issued to both Singaporean and Hong Kong universities represent relatively high-quality patents for these non-US universities.

    The data on USPTO patenting by Hong Kong and Singaporean universities were collected from the USPTO website (http://patft.uspto.gov/netahtml/PTO/search-bool.html; accessed 15 September 2008) by searching the "assignee" field for university names. It is possible, therefore, that patents initially not assigned to these universities are not included in the tabulation, or that various misspellings and typographical errors in the USPTO assignee data may cause the omission of observations. The crude methods used to construct this dataset provide another set of reasons to interpret findings from this discussion with caution.
20. Because this measure is based on the application date for issued patents, and review of patent applications can take as long as 3–5 years, the data are truncated, producing a decline in the number of applications for issued patents after 2002.

21. The 2007 *China Science and Technology Statistics Data Book*, published by the PRC Ministry of Science and Technology, reports that central and local government expenditures on R&D rose from slightly more than 70 billion yuan in 2001 to nearly 169 billion yuan in 2006 (http://www.most.gov.cn/eng/statistics/2007/200801/P020080109573867344872.pdf; accessed 23 October 2008). These reported expenditures do not include R&D funded by state-owned enterprises.
22. Nevertheless, Sharif and Baark (2006) cite an interview with the CEO of an Hong Kong microdisplay firm who criticized the HKUST Technology Transfer Center (TTC) for focusing too narrowly on building its patent portfolio and providing minimal assistance to his firm: "Having drawn on the TTC's services in the past, the microdisplay manufacturer now prefers to go it alone, engaging with the TTC at only a superficial level to gain access to HKUST resources such as laboratories and equipment, and for proper documentation of practices and procedures" (p. 19). It is unclear whether the experience of this one firm is representative of Hong Kong firms' dealings with HKUST more generally.
23. I did not attempt to obtain licensing data from the universities that we visited, since these data generally are sensitive and available only under nondisclosure agreements, if then. Licensing data (identifying the licensee and the extent of any royalty income) would allow an evaluation of the statements from most interviewees that Hong Kong universities had negotiated relatively few licenses with non-Hong Kong enterprises, such as multinationals (HKU mentioned licensing biomedical technologies to some multinational firms).
24. Analysis of licensing data, including the identity of licensees, would enable this characterization to be assessed.
25. "In the Public Interest: Nine Principles to Consider in Licensing University Technology," 6 March 2007. Signatory universities and organizations were Caltech, Cornell, Harvard, MIT, Stanford, the University of California, the Wisconsin Alumni Research Foundation, the University of Illinois-Chicago, the University of Illinois-Urbana Champaign, Yale University, the University of Washington, and the Association of American Medical Colleges.
26. One exception to this general characterization is the Technology Cooperation Funding Scheme (TCFS), which is part of the ITF's Innovation and Technology Support Program. The TCFS supports collaborative R&D projects involving Hong Kong universities or firms and firms of Guangdong province, and requires that firms provide a portion of the funding for projects. Although the R&D Centers within Hong Kong are eligible recipients of funding through this program, there is little other evidence suggesting that the TCFS's R&D grants are focused on specific areas of R&D and/or that they are coordinated with funding programs overseen by the Guangdong provincial or the PRC central governments (see http://www.itf.gov.hk/eng/TCFS.asp for additional details).
27. Moreover, new, knowledge-intensive firms may spin off from sources other than universities, as Klepper (2008) has pointed out. In particular, established firms in high-technology industries are an important source of new firms. Policy-related obstacles to the formation of such spin-offs include "noncompete" agreements and other limitations on employee mobility.

HKSAR policymakers wishing to encourage entrepreneurship should consider a review of these policy-related obstacles to the "spawning" of new firms by established firms with a view to removing them where possible. A fuller discussion of the importance of this phenomenon and the extent of these obstacles to new-firm formation in Hong Kong unfortunately is beyond the scope of this paper.

References

Association of University Technology Managers (AUTM) (2001) *The AUTM Licensing Survey: FY 2000*. Association of University Technology Managers. Norwalk, CT: AUTM.

Association of University Technology Managers (AUTM) (2002) *The AUTM Licensing Survey: FY 2001*. Association of University Technology Managers. Norwalk, CT: AUTM.

Baark, E. (2006) "The Development of Innovative Capacity in Hong Kong," presented at the conference on "Greater China's Innovative Capacities: Progress and Challenges," Tsinghua University, Beijing, 20–21 May.

Berger, S., and R. Lester (1997) *Made by Hong Kong* (Hong Kong: Oxford University Press).

Census and Statistics Department (HKSAR) (2007) *Report on 2006 Annual Survey of Innovation Activities in the Business Sector* (Hong Kong).

Census and Statistics Department (HKSAR) (2008) *Report on 2007 Annual Survey of Innovation Activities in the Business Sector* (Hong Kong).

Census and Statistics Department (HKSAR) (2008) "Statistics on Research and Development of Hong Kong, 2002 to 2006,"*Hong Kong Monthly Digest of Statistics*, May.

Chinese University of Hong Kong (2008) "Facts and Figures 2008: Research" (http://www.cuhk.edu.hk/iso/facts/issue/2008/research_e.htm).

Cohen and Levinthal (1990) "Absorptive Capacity: A New Perspective on Learning and Innovation," *Administrative Science Quarterly* 35, 128–152.

Crespi, G.A., A. Geuna, and L.J.J. Nesta (2006) "Labour Mobility of Inventors: Career Decision and Knowledge Transfer," working paper, European University Institute.

Gulbrandsen, M., and L. Nerdrum (2009) "University-Industry Relations in Norway," in J. Fagerberg, D.C. Mowery, and B. Verspagen, eds., *Innovation, Path Dependency and Policy: The Norwegian Innovation System* (Oxford, UK: Oxford University Press).

Hall, B.H., A.B. Jaffe, M. Trajtenberg (2001) "The NBER patent citations data file: Lessons, insights and methodological tools," NBER Working Paper 8498, Cambridge, MA. "In the Public Interest: Nine Principles to Consider in Licensing University Technology," 2007. Stanford University Office of Technology Licensing; processed. 6 March.

Innovation and Technology Fund of Hong Kong (2008) "Statistics of Approved Projects" (http://www.itf.gov.hk/eng/statistics/StatTable101ew.asp?StatTypeId=101&StatId=516&StatCaption=Statistics+of+Approved+Projects).

Innovation and Technology Commission of Hong Kong (2007) "A Consultation Paper on the Proposed Relaxation Measures of University-Industry Collaboration Programme under the Innovation and Technology Fund" (http://www.itf.gov.hk/eng/Forms/Consultation_Paper_on_UICP.pdf).

Innovation and Technology Commission of Hong Kong (2008) "Guide to the Innovation and Technology Fund University-Industry Collaboration Programme" (http://www.itf.gov.hk/eng/Forms/itf-uicp-guide.pdf).

Innovation and Technology Commission of Hong Kong (2008) "Guangdong-Hong Kong Technology Cooperation Funding Scheme (TCFS)" (http://www.itf.gov.hk/eng/TCFS.asp).

Interview, Cliff Chan, Tele Eye, Hong Kong, 10 June 2008.

Jiao Tong University, "Academic Ranking of World Universities," Shanghai, China (http://www.arwu.org/ARWU2007.jsp; accessed 12 February 2008).

Klepper, S. (2008) "Silicon Valley – A Chip Off the Old Detroit Bloc," in D. Audretsch and R. Strom, eds., *Entrepreneurship, Growth, and Public Policy* (New York: Cambridge University Press).

Ministry of Science and Technology, People's Republic of China (2008) China Science and Technology Statistics Data Book (http://www.most.gov.cn/eng/statistics/2007/200801/P020080109573867344872.pdf; accessed 15 September 2008.).

Mowery, D.C., B.N. Sampat, and A.A. Ziedonis (2002) "Learning to Patent: Institutional Experience and the Quality of University Patents," *Management Science* 48, 73–89.

OECD (2007) *OECD Science, Technology, and Industry Outlook 2006*.

Research Grants Council of Hong Kong (2008) *Annual Report 2007* (Hong Kong: University Grants Committee).

Sharif, N., and E. Baark (2006) "Mobilizing Technology Transfer from Universities to Industry: The Experience of Hong Kong Universities," unpublished MS.

U.S. Patent and Trademark Office, "Patent Full-Text Image and Database" (http://patft.uspto.gov/netahtml/PTO/search-bool.html).

University Grants Committee of Hong Kong (2008) "Facts and Figures: Academic Year 2007/2008" (http://www.ugc.edu.hk/eng/ugc/publication/report/figure2007/15.htm).

University Grants Committee of Hong Kong (2008) "Applications for Allocations from the General Research Fund 2008/2009" (http://www.ugc.edu.hk/eng/rgc/grf/application/cergia.htm).

University Grants Committee of Hong Kong (2008) "Research Grants Council General Research Fund 2010/2011: Explanatory Notes (GRF 2) for Completing the Application Form" (http://www.ugc.edu.hk/eng/doc/rgc/form/GRF2.pdf).

2
Hong Kong's Intellectual Property Rights Regime and Innovation Policy

Poh-Kam Wong
National University of Singapore

1. Introduction

This chapter seeks to complement the various sector-specific chapters in this project by examining the possible role of public policy on intellectual property (IP) in enhancing the Hong Kong Special Administrative Region's (HKSAR) innovation system. While there are various interpretations of the scope of IP policy in the literature (see e.g., Idris, 2003; JSCIP, 2002; UNIDO, 2006), in this chapter we take a broad interpretation that covers not only policies to protect and enforce IP rights, but also policies to promote the creation of IP and facilitate its commercial exploitation and market transactions. In addition, besides examining the efficacy of the enacted policy framework, we also look into the effectiveness of institutions in policy implementation and enforcement.

Prior work on Hong Kong's economic competitiveness has offered contrasting views on the possible role of IP policy in Hong Kong. On the one hand, authors like Enright et al. (1997) have argued for a minimalist state role. They point out that a major pillar of Hong Kong's past phenomenal success is its legal system characterized by the strict rule of law and judicial independence. However, in contrast to other developmental states in East Asia like Japan, Korea, Singapore, and China, the unique strength of Hong Kong's economic system, they argue, has been its laissez-faire approach, popularly described as "positive non-intervention," whereby the government provides the necessary legal framework to facilitate free trade and efficient market transactions, but otherwise does not intervene in favor of any particular industry, nor does it directly engage in economic activities through state-owned or state-controlled enterprises. In this minimalist state perspective, the government of Hong Kong should continue its strong focus on protecting property rights in general, and intellectual property rights in particular. However, the state should have little role in promoting IP creation and commercialization, relying instead on market forces to determine the level of IP creation and usage.

In contrast, Berger and Lester (1997), in their work *Made by Hong Kong*, have highlighted the need for the state to play a stronger role with respect to IP. Besides pointing out that, as of the mid-1990s, Hong Kong's record of enforcement of IP rights has been patchy, they make a strong case for the government to take a more proactive role in promoting a culture of IP creation among the local industries. It is interesting to note that, while Enright et al. (1997) do not include any entry on Intellectual Property in their index, Berger and Lester (1997) make numerous references to IP in their index.

In this chapter, I revisit the above debate by first examining the salient developments of both aspects of IP policy – IP protection and IP creation/commercialization promotion – in HKSAR in the period after the end of British colonial rule in 1997. I then examine international comparative evidence on HKSAR's IPR protection regimes versus selected reference countries over the years, as well as providing a comparative analysis of the pattern and trend of patenting over the last two decades between HKSAR and Singapore, a developmental state that has adopted a more interventionist policy approach towards IP. Based on the above comparative analysis, I offer a number of recommendations on the role of IP policy in the future development of HKSAR's innovation system.[1]

2. Development of IPR protection policies in HKSAR

By international standards, HKSAR has a relatively well-developed legal framework for protecting intellectual property rights (IPR). Hong Kong's mini-constitution – the Basic Law – specifically provided in Articles 139 and 140 that HKSAR should, on its own, develop appropriate policies and afford legal protection of IPR. An IP Department (IPD) was established in July 1990 with three stated objectives: (a) to advise the Secretary of Commerce, Industry and Technology on policies and legislation to protect IP in the HKSAR; (b) to operate HKSAR's trade marks, patents, registered designs and copyright licensing bodies' registries; and (c) to promote IP protection through public education (IP Department, 2004).

With the return of Hong Kong to China and its formal designation as HKSAR in 1997, the various IP-related legislations prevailing in Hong Kong were updated. The new Patents Ordinance, Copyright Ordinance and Registered Design Ordinance came into effect in June 1997, while a revised Trade Marks Ordinance (Cap 559) was introduced in April 2003. As part of China, HKSAR has also been covered by all the major international IP conventions, including the Paris Convention, the Berne Convention, the Universal Copyright Convention, and the Patent Cooperation Treaty (PCT). HKSAR also became a member of the World Trade Organization WTO in its own right, and signed the WTO Trade-related aspects of intellectual property rights (TRIPS) Agreement.

It should be noted that, notwithstanding the introduction of its own IP legislation, HKSAR post-1997 continues to operate a patent system based on "re-registration" of patents registered elsewhere, rather than having its own system of patentability examination. During colonial times, patent protection in Hong Kong was provided by re-registering a patent filed with the UK Patent Office, and the protection lasted as long as the original patent in the UK. Although this registration system was later extended to cover patents filed with the European Patent Office (EPO) and the PRC State Intellectual Property Office (SIPO) under the new IP legislation, HKSAR has not as yet introduced its own examination system. HKSAR's IPD did introduce a short-term patent ("petty patent") system of its own that provides protection of shorter duration (a maximum of 8 years compared with 20 years for standard patents) for inventions with limited novelty that may otherwise not qualify for full patent protection, and for which prior registration elsewhere is not required. However, the examination system set up by IPD for such short-term patents is rather limited in scope, and mainly entails submission of a search report by the UK Patent Office, EPO and SIPO or a PCT-recognized international searching authority.

Besides having a relatively updated and comprehensive IP legislative framework since 1997, HKSAR has also made significant progress in terms of actual enforcement of IPR protection, as suggested by data from the IP Department. As late as 1997, Hong Kong was still placed on the 301 Watch List of the US Trade Representative. However, with the enactment of the Prevention of Copyright Piracy Ordinance and amendments to the Import and Export Ordinance in 1997, the Customs and Excise Department (CED) set up a special task force to implement more rigorous enforcement actions, resulting in a drastic drop in the reported incidence of copyright piracy, particularly pirated optical discs.

By 1999, HKSAR had been removed from the 301 Watch List (www.ipd.gov.hk/eng). More recently, HKSAR has also stepped up enforcement of trademark violation cases (Lam, 2006).

The IPD has embarked on a series of public educational campaigns in recent years to raise awareness of IP and IPR protection. Consequently, the department was able to report in a survey it conducted in 2003 that over 90 percent of HKSAR residents agreed that IPR protection was necessary; this increased to 96.3 percent in its latest survey in 2008. In addition, the IP Department has made substantial efforts to streamline the administrative processing of IP applications and registration, including the provision of electronic searching, filing, payment and publication services for patents, copyrights and designs since 2003. In 2006, new interactive services were launched that significantly improved the response time for various IP administrative processes, such as a change of particulars of IP owners.

The improvement in the IPR protection regime of HKSAR since 1997, as reported by the IP Department, is borne out by a number of international comparative studies that provide indices for measuring IPR protection across countries, including the *Economic Freedom of the World Annual Report* and the annual *Global Competitiveness Report* (GCR). As can be seen from Table 2.1 below, HKSAR's overall patent rights index rose only slightly from 2.46 in the late 1970s to 2.57 in 1995, but increased at a higher rate to 2.90 in 2000.

Table 2.1 Patent rights index, *Economic Freedom of the World Report*

	1960–1975	1975–1980	1995	2000
Hong Kong	2.04	2.46	2.57	2.90
China	n.a.	n.a.	1.55	2.48
India	1.68	1.57	1.51	2.18
Japan	3.24	3.94	3.94	4.19
Korea	2.87	3.61	4.20	4.20
Singapore	2.37	2.57	3.90	4.05
US	3.86	4.41	4.86	5.00
Switzerland	2.84	3.80	3.91	4.05
Ireland	2.69	2.99	3.32	4.00

Note: The index is based on five categories: (1) coverage (the subject matter that can be patented); (2) duration (the length of protection); (3) enforcement (the mechanisms for enforcing patent rights); (4) membership in international patent treaties; and (5) restrictions or limitations on the use of patent rights.

Source: 1960–75, 1975–90: W. G. Park, "Intellectual Property & Patent Regimes," *Economic Freedom of the World: 2001 Annual Report*, chapter 4; 2000: W. G. Park & S. Wagh, "Index of Patent Rights," *Economic Freedom of the World: 2002 Annual Report*, chapter 2.

Notwithstanding the progress made, HKSAR's overall patent rights index in 2000 remained significantly below those of Japan and Korea, which had achieved among the highest level of patenting outputs in Asia, along with Singapore and Ireland, which had registered significant growth in patenting since the mid-1990s.

More recent comparisons are unfortunately not available, as the *Economic Freedom of the World Report* has switched to adopting the IPR protection index provided by the GCR after 2000. Unlike the Patent Rights Index, GCR's IPR protection index covers all aspects of IP, not just patents, and is derived from a survey of business executives using a Likert scale. As Table 2.2 shows, HKSAR appears to score relatively high based on this broader index, ranking it higher than Korea and Taiwan. However, it remains below the level achieved by Japan, Singapore and Ireland. Moreover, while the index for HKSAR has stayed flat between 2003 and 2008, those of Japan, Korea, Singapore and Ireland have continued to improve.

The overall evidence thus appears to suggest that HKSAR has seen improvement in its IPR protection legal framework and enforcement institution since 1997, but that it remains below a number of other economies that either have had a longer history of promoting IP (particularly Japan and Switzerland) or that have emphasized the adoption of

Table 2.2 GCR IP rights protection index, 2000–2008, selected years

	2000 (out of 10)	2003 (out of 7)	2008 (out of 7)
HK	6.3 (4.4)	5.3	5.4
China	3.22 (2.3)	3.4	3.9
India	3.27 (2.3)	3.5	3.7
Japan	7.55 (5.3)	4.7	5.7
Korea	5 (3.5)	4.5	5.0
Singapore	7.62 (5.3)	5.9	6.3
Taiwan	n.a.	5.0	4.9
US	9.1 (6.4)	6.2	5.6
Switzerland	9.17 (6.4)	5.9	6.3
Ireland	7 (4.9)	4.7	5.6

Notes: In 2000, the Likert scale is 1–10. Figures in brackets are re-scaled to the 1–7 range.

In 2003, the index is measured by responses to the following: IP protection in your country is (1 = weak and non-existent, 7 = equal to the world's most stringent).

The index in 2008 is measured by responses to the following: Intellectual property protection and anti-counterfeiting measures in your country are (1 = weak and not enforced, 7 = strong and enforced).

Source: *Global Competitiveness Report*, various years.

business-friendly policies to attract international inward direct foreign investment (DFI) (Singapore and Ireland).

3. Development of IP creation and commercialization policies in HKSAR

Besides ensuring the protection and enforcement of IPR, another major IP-related public policy thrust pertains to the promotion of IP creation and its commercial exploitation. While a major driver for IP creation is the level and intensity of involvement of public research institutions, enterprises and individuals in R&D and other creative activities, experience in newly industrialized economies suggests that there may be an additional role for the state to explicitly promote the translation of R&D efforts into tangible intellectual property, over and above the policies to promote R&D and innovation activities themselves (see e.g., JSCIP, 2002). For example, patenting output of Taiwan's Industrial Technology Research Institute (ITRI) rose substantially from the mid-1980s, after a change in the strategic direction of the institute leadership to emphasize patenting output as a performance indicator (private communication, C. C. Lin, ex-president of ITRI).

Promotional policies for IP creation can take various forms, for example through the use of financial subsidies to offset the cost of applying for IP protection, or the use of IP creation and its subsequent commercial exploitation as performance indicators (e.g., licensing income) to incentivize public research institutions to create and commercialize IP. High-prestige public recognition of IP success in the form of national awards may similarly incentivize private enterprises and individuals to focus efforts on IP creation and commercialization *outputs*, over and above the award of *input* resources to promote R&D activities such as public R&D grants and innovation subsidies.

Besides supply-side policy, public policies can also promote the commercial exploitation of IP by stimulating market demand and improving the efficiency of the market transaction process. Market demand can be boosted by financial incentives for enterprises, particularly small-medium enterprises (SMEs), to adopt new technology through in-licensing of IP. Finally, to the extent that the market for IP transactions is imperfect, there may be a role for government intervention to facilitate the development of the IP transaction markets through various means, for example by stimulating the development of the intermediary IP professional services industry, subsidizing the training of IP professionals, facilitating the formation of IP-related industry associations, and providing public funding for infrastructure that facilitates IP transactions (e.g., an online IP database and IP market exchange). These public policies should be understood as temporary in nature, to fill a temporary gap in market development, or to jump-start/speed up the intermediary industry development process itself.

Because the creation and exploitation of IP is intimately linked to the innovation process in any innovation system, it is obviously difficult to completely isolate policies designed to stimulate IP creation and commercialization from general policies to promote R&D and innovation activities. Indeed, even the IPR protection policies described earlier are ultimately meant to have a direct incentivizing impact on IP creation and commercialization. It is also likely that some of the IP promotion policies are implemented by the very same organizations in charge of general promotion of R&D and innovation activities. Nevertheless, by examining the various public institutions involved in R&D and innovation promotion, one can get a sense of the extent to which these institutions and their programs have explicitly incorporated policy elements aimed at stimulating IP creation and commercialization. In addition, by examining the state of development of the IP professional services industry, one can gauge the maturity of the market for IP commercialization, and hence the potential need for public intervention.

3.1. Strategic IP policy directions by the Innovation and Technology Commission

Any assessment of HKSAR's public intervention in innovation promotion must start with the Innovation and Technology Commission (ITC), which was established in July 2000 with the explicit mission to spearhead HKSAR's drive to become a world-class, knowledge-based economy. To achieve its mission, the ITC functions primarily as a funding agency that promotes specific innovation activities through various funding schemes (ITF, ARF and DesignSmart), and as a developer of various infrastructural facilities and institutions (HKSTPC, ASTRI, HKPC, HKDC, etc.). As the specific goals and scope of the functions of these ITC-funded schemes and institutions have already been analyzed elsewhere in this project, I will focus instead on examining the possible roles that ITC is playing to promote IP creation and exploitation through these funding schemes and institutions.

Based on my reading of various ITC documents and interviews with selected senior officials at ITC, my impression is that, while ITC's overall objectives implicitly cover the promotion of IP creation as part of its mission of promoting innovation, it does not appear to have a high-level, explicit policy targeted at increasing IP output per se, beyond stimulating R&D and innovative activities in general. ITC does provide a patent application grant that subsidizes the cost of patent application by HKSAR residents, but the reach and impact of this grant scheme appears to be modest. With regard to IP management, ITC has a policy of allowing universities and R&D Centres to retain ownership of IP created under platform projects. It also provides R&D Centres with guidelines on handling IP sharing arrangements under platform and collaborative projects. Beyond this, ITC does not appear to have set any specific policy guidelines on how the IP that is created through its funding should be managed. For example, there appears to be no explicit guidelines imposed by ITC on how the Applied Science and Technology Research Institute (ASTRI) and the various universities should manage the IP that is created from ITC funding – for example whether they should have 100 percent ownership of the IP, the licensing policy of the IP, how the royalties generated from the licensing of the IP are to be distributed between the owning institution and inventors, etc. Instead, with some notable exceptions, the institutions receiving the ITC funding appear to have the autonomy to decide how to manage their IP. Such a hands-off approach has the advantage of allowing the funded institutions the flexibility to fashion their own IP policies. However, it also has the disadvantage of a potential loss of strategic influence.

A good example of this autonomy policy is that of ASTRI, which was set up to perform applied R&D in selected high-tech fields with the aim of transferring the results to industry for commercialization. In recent years, ASTRI has been given the autonomy to experiment with alternative mechanisms for IP commercialization, including spin-offs in the early years, and to focus on IP licensing. Discussion with senior management of ASTRI suggests that the criteria for measuring the performance of the institute have been evolving, and there has been no direct, long-term strategic top-down policy imposed by ITC. Thus, while ASTRI management has paid greater attention to licensing income generation in recent years, this is only in response to perceived future ITC funding constraints, and not a direct IP creation policy imposed by ITC.

The most notable exception, where ITC imposes an explicit IP policy through its funding, is in the granting of the University–Industry Collaboration Programme (UICP) funding to private enterprises that wish to leverage the expertise of Hong Kong universities in their R&D activities. Provided that the companies are contributing at least 50 percent of the R&D costs, with a UICP grant providing the balance, all IPR generated from these R&D projects will accrue to the companies. While it is intended to incentivize private industry to tap university expertise, the strong IP rights that UICP grants to private industry may discourage university faculty with cutting-edge research capabilities to engaging in such activities.

3.2. IP policy in the public university sector

Besides the ITC, the public university sector plays a significant role in the innovation system of HKSAR. As highlighted by David Mowery in his chapter on university–industry collaboration and technology transfer in HKSAR for this project, the five major public universities in HKSAR, despite being funded in similar ways by the government, pursue quite diverse approaches to technology commercialization in general and IP management in particular. Moreover, their strategies have changed during the last 20 years, with some cutting back on direct involvement in technology commercialization promotional activities, and others expanding the scope of involvement through, for example, the establishment of venture capital funds. In their study of HKUST, Sharif and Baark (2009) described the significant changes in the university's IP policy over time, moving from focusing on spin-offs in the 1999–2001 dot-com boom period to a greater emphasis on licensing from 2003 onwards. Leung (2008) also highlighted the diverse performance of the eight public universities in terms of spin-offs that exploit IP generated by university research.

The relative autonomy of the individual universities to pursue their own technology transfer and IP management strategies reflects the lack of strategic top-down direction from the HKSAR government. As highlighted by Mowery (this volume), this lack of strategic policy direction by the Hong Kong government extends beyond the link between universities and industry in HKSAR itself, to links between those universities in Hong Kong with innovation activities in Mainland China and universities and industry there.

3.3. Development of the IP professional services industry

While governments in a number of newly industrialized economies, notably Singapore and Ireland, have actively been promoting the development of various IP professional services in their economies, a similar focus appears to be lacking in HKSAR. To begin with, the ITC does not seem to regard as part of its mission to promote the development of the IP services industry, neither does it see itself as playing a role in promoting the training and development of IP professionals.

Likewise, the IP Department (IPD) appears to be focused primarily on promoting awareness of IP rights and on formulating and implementing policies to effectively enforce the protection of IP rights. As highlighted earlier, while the IPD appears to have done a commendable job in terms of conducting regular mass media publicity campaigns, keeping abreast of and up to date on the latest IP protection legislation worldwide, making the process of registering and examining IP applications efficient, and carrying out high visibility enforcement exercises, it does not seem to have a mandate on promoting the development of the IP professional services industry as a means to stimulate IP commercialization.

HKSAR does have a government agency, Invest Hong Kong, charged with promoting investment. However, discussion with senior officials at Invest Hong Kong suggests that, while the agency's overall mandate is to make Hong Kong a desirable place for local and foreign businesses to invest and operate in, its primary policy tool is to keep taxes low and public infrastructure efficient, and to provide a transparent legal and business-friendly environment. As such, Invest Hong Kong does not pursue any industrial policy in the sense of providing special incentives to particular industries. The development of the IP professional services industry is seen to be best left to market forces, and not for the government to promote.

There are no reliable statistics on the size and sophistication of the IP professional services industry in HKSAR. A 2004 study by Jinan University, commissioned by the IPD, estimated the number of certified attorneys engaging mainly in IP-related practices in HKSAR to be around 100, based on membership of the Hong Kong Institute of Trade Mark Practitioners and the Hong Kong chapter of APPAA (Asia-Pacific Patent Attorney Association) (Jinan University, 2004). However, the same study highlighted that work related to trademarks forms the largest category of services provided, with patenting of lower significance. Moreover, my interview with the senior partners of a couple of leading private law firms with significant IP practices, who are executive committee members of the Hong Kong chapter of APPAA, suggested that the actual number of law firms with significant IP practice (more than one full-time lawyer equivalent) is small, probably less than 10. They also highlighted a concern that there is no requirement for certification or qualification of trademark agents in Hong Kong, and as a result, quality and professional standards have been rather inconsistent. Lobbying by the APPAA for such certification has been rejected by the IPD, out of concern that this represents a restriction of market access by the larger law firms. In addition, the interviews suggested that much of the patenting work in HKSAR itself is of an agency nature, with the substantive work of patent drafting and prosecution strategy actually contracted to patent specialists in the US or UK. Confirming the finding of the Jinan University study, IP litigation and IP strategic consulting work were also cited as being of lesser importance.

The Jinan University study also found that the HKSAR IP professional services firms tend to serve largely customers in HKSAR itself, with relatively little reach to customers in Mainland China and the Pearl River Delta (PRD) regional hinterland. Despite the signing of the Closer Economic Partnership Agreement (CEPA) between Mainland China and HKSAR, only one quarter of the IP professional services firms in HKSAR surveyed by Jinan University reported any intention to expand operation to Mainland China in the foreseeable future. The survey also found only a small proportion of clients of IP professional services in Guangdong Province actually engaged the services of HKSAR-based firms. My interviews with several IP professional services firms in Beijing and Shanghai similarly suggest that few clients in these two cities use the services of HKSAR-based firms.

It is not clear whether the lack of a local patent examination system in HKSAR may have hindered the development of its IP professional services industry, although the experience of other jurisdictions that have seen rapid growth in the IP professional services industry, including Taiwan and China, seems to suggest that this could be the case. Arguably, the establishment of a local examination system (in contrast with the current registration system) would raise demand for local IP expertise, and, as evidence from other countries would suggest, the IP examination system has often been a training ground for the manpower who staff the private IP professional services industry. A local examination system would also lead to development of specialization in the court system to handle IP litigations cases and to the establishment of its own case laws over time.

It is also unclear whether HKSAR's adoption of an alternative short-term patent system has facilitated or hindered the development of its IP professional services industry. While more than 20 countries around the world have implemented some form of "petty patent" system to provide protection innovation that has a lower standard of inventiveness, the experience has been mixed. In HKSAR, this has been used primarily to protect toys and electronic goods. Due to their limited inventiveness requirements, they are more likely to have spurred the growth of design services (see below) rather than advanced IP expertise.

3.4. Policy to facilitate IP exploitation by local SMEs

In many countries, a major focus of public innovation policy concerns the lack of capacity by local SMEs to exploit IP created by public research institutes, universities or other private enterprises. Besides the lack of technical knowledge and financial resources, many local SMEs are also not familiar with the process of technology transfer and the complexities of IP licensing. Policies to facilitate IP exploitation by local SMEs can take various forms, for example directing public research institutes to license their IP cheaply to local SMEs, or giving subsidy to the local SMEs to reduce their cost of licensing external technologies to improve their business.

In the case of HKSAR, the ITC directly funds a number of public applied R&D Centres to conduct applied R&D and provide technical assistance to local SMEs in a number of manufacturing industries (textile and clothing, logistic and supply chain, automotive parts and components, etc.), but the emphasis appears to be on fostering joint R&D and technical support rather than on developing IP portfolios to be licensed cheaply to local enterprises. ASTRI does appear to have a strong focus on developing a portfolio of IP for licensing, particularly in semiconductor chip design technology (see Fuller, Chapter 10 of this volume). However, there appears to be a mismatch with the needs of local SMEs in HKSAR, as many of the enterprises that are keen to license ASTRI's technologies appear to be based in China.

There seems to be no program that provides a subsidy to the local SMEs themselves to encourage them to adopt new innovation (through in-licensing of new technologies), although ITC does run a New Technology Training program that provides subsidies for the training of manpower in new technical skills.

3.5. Development of industrial design capability

According to registration statistics produced by the IPD, industrial design represents a significant form of IP creation in Hong Kong. My interviews with a number of indigenous design firms and a former manager at the Hong Kong Design Centre indicate some governmental efforts in promoting the design industry. However, a major criticism was raised that the primary focus had been on physical infrastructure (a building to house design firms), with inadequate attention paid to the development of indigenous design skills and the nurturing of design business and IP management capabilities among local design firms. There was also criticism that local university industrial design courses did not appear to include sufficient exposure to the business and IP aspects of industrial design.

3.6. Development of early-stage venture capital and angel investment groups

Besides the IP professional services firms, a related industry that has been found to be critical in supporting technology commercialization through start-ups is that of venture capital firms and angel investors who have the technology and IP know-how to invest in early stage, IP-based, high-tech start-ups (see e.g., Wright et al., 2006). Interviews with the Hong Kong Venture Capital Association suggest that the venture capital industry in HKSAR is dominated primarily by late-stage venture capital firms and private equity funds, with very little early stage venture capital funding available. Moreover, while individual angel investors do exist who invest in early stage start-ups, there appears to be no formal business angel networking groups or associations (like the Band of Angels in Silicon Valley) that organize regular networking activities to match start-ups with investors.

The Cyberport project was originally conceived as a platform for incubating start-ups in ICT. However, in the aftermath of the dot-com crash in early 2000, the project appeared to have evolved into a primarily physical infrastructure project, with relatively little emphasis on additional value-adding activities to enhance IP commercialization capability. There was some attempt to involve mentoring by senior industry players, and to provide networking with potential angel investors, but the scale of the efforts appears to have been quite modest. Moreover, its distance from the public universities is a deterrent to close interactions with university students and researchers.

In summary, in contrast to the formulation and implementation of IPR protection policies, which have received high-level attention, with considerable new legislative updates and institutional enforcement efforts to stay abreast with international developments, public policies to intervene directly in promoting IP creation and commercialization appear to be lacking strategic direction from the top, and are generally implemented in a more diffused, ad hoc manner. In particular, there seems to be a lack of strategic policy focus on promoting HKSAR's IP professional services industry.

4. IP creation pattern and trends in HKSAR compared with Singapore since 1997

To what extent has the apparent lack of strategic focus on IP policy by the Hong Kong government affected the pace of IP creation and commercialization in HKSAR since 1997? While there are many factors that influence the rate of IP creation and exploitation, it would nonetheless be useful to examine the actual pattern and trend of IP creation in HKSAR over time, as benchmarked against Singapore, which arguably has adopted a more interventionist approach to promoting IP creation. For this comparative analysis, we will focus primarily on patenting rather than on all classes of IP, given the closer link of patents to technological innovation activities. We use, in particular, data on utility patents granted by the United States Patents and Trademark Office (USPTO) as the basis for comparison, due to the lack of comparability of national-level patenting records that may reflect differences in ease of obtaining patent protection, but also because the commercial importance of the US market provides a better gauge of the commercial potential of the patents (Jaffe and Trajtenberg, 2002).

We believe that such a comparative analysis is more instructive than looking at patenting output trends in HKSAR alone, given that IP creation tends to increase over time in virtually any growing economy. While both HKSAR and Singapore have developed relatively efficient systems for IPR protection, the aim of the comparative analysis is to discern if the rate of growth in IP creation is higher in Singapore with its more interventionist role in promoting IP creation.

Tables A.1–A.9 and Figures A.1–A.2 provide more detailed information on the trends and patterns of patenting in HKSAR compared with Singapore. Collectively, they portray considerable differences in the trend and pattern of patenting in the two economies over the last three decades. We highlight in particular the following salient differences:

i) While HKSAR and Singapore have both experienced rapid growth in US patenting, Singapore's growth rate has been faster since the mid-1980s (Table A.1).
ii) The difference became even more marked when we excluded design patents and only examined utility patents (Table A.2).

iii) Since the early 2000s, Singapore has overtaken HKSAR in terms of number of US utility patents; on a per capita basis, the differences in invention patenting intensity between Singapore and HKSAR is even more marked (Table A.3 and Figure A.1). Even if we adjust for the lower R&D expenditure per capita in Hong Kong compared with Singapore, Singapore remains ahead in recent years.

iv) In terms of trademark registration, while HKSAR continued to have a higher number of trademark registrations than Singapore, the latter experienced a higher overall growth rate (Table A.4).

v) HKSAR's patenting is characterized by a rather high proportion of design patents, compared with not only Singapore, but other countries like Japan, Korea and the USA (Table A.5).

vi) A larger proportion (72 percent) of HKSAR's patents are owned by local assignees than is the case with Singapore, where half of the patent inventions are owned by foreign entities, primarily foreign multinational corporations (MNCs) with R&D operations in Singapore (Table A.6). It is no surprise that the 20 largest patent-owning organizations in Singapore are dominated by foreign firms (Table A.9b), whereas in HKSAR, local organizations have a bigger presence than foreign firms (Table A.9a).

vii) HKSAR's patenting is also characterized by a relatively higher proportion of ownership by individuals than by organizational assignees, compared not only with Singapore (Table A.6), but also other countries like Japan, Korea and the USA.

viii) Universities in HKSAR appear to contribute a smaller share of US patenting than universities in Singapore in their respective economies (2.7 percent compared with 4.4 percent). Likewise, the share of patenting by public research institutes in HKSAR is negligible, but constituted 5.5 percent of total patenting output in Singapore (Table A.6).

ix) Using forward citations as a proxy measure of patent quality, HKSAR's patents are found to be of lower quality than those of Singapore since the mid-1980s. This finding is true whether we use the average forward citations per patent as the quality indicator (Table A.7a), the relative citation index (Table A.7b) or the share of high-impact patents (Table A.7c).

x) While electrical and electronics technologies have become the largest technology field for patenting in both HKSAR and Singapore since the mid-1980s, the specialization in this field has become more marked in Singapore (close to 50 percent in the last 10 years). Moreover, computers and communication technologies have also become relatively more important in Singapore than in HKSAR in recent years (Figure A.2). As a result, the degree of concentration by technological fields has increased in Singapore over the last 10 years, while that in HKSAR has declined (Table A.8).

Overall, the statistical evidence from US patenting seems to suggest that HKSAR has lagged behind Singapore in recent years in terms of both the quantity and quality of utility patent creation. While HKSAR continues to lead in design patent and trademark registration, Singapore is registering higher trademark growth rate. Singapore's greater degree of technology specialization, and its greater contribution of public research institutes and universities in patenting output, is reflective of the greater role of the public sector in the national innovation system of Singapore compared with HKSAR, in terms of both the relative share of innovation activities conducted by the public universities and public research institutes, as well as the deliberate DFI policy to attract foreign MNCs to conduct R&D in targeted technology fields in Singapore.

The above profile of patenting outputs in HKSAR and Singapore is also consistent with findings on the pattern of R&D and innovation activities as revealed by R&D surveys in both economies. Based on the 2007 survey of innovation activities in the business sector in Hong Kong (Census and Statistics Department, HKSAR, 2008), the largest contributor (45 percent) of in-house R&D expenditure in HKSAR is the commerce and trades sector, followed by finance and business services (36 percent), with the manufacturing sector contributing less than 10 percent. In contrast, according to the 2006 R&D survey of Singapore, manufacturing accounted for 67 percent of private industry R&D. While small and medium enterprises accounted for more than half of all business R&D expenditure in HKSAR, large manufacturing enterprises, primarily foreign MNCs, contributed over two-thirds of private sector R&D spending in Singapore.

Similar findings can be derived using utility patenting data from the European Patent Office (EPO). For example, in the period between 1976 and 1997, Hong Kong had more EPO utility patents than Singapore (133 compared with 112), but in the subsequent decade (1998–2007), HKSAR's EPO utility patent counts had fallen to only one-third of those of Singapore (182 compared with 529).

While it is difficult to ascertain the extent to which the observed differences in IP creation rate can be attributed to the more proactive role of the state in Singapore in promoting IP creation than in Hong Kong, it seems plausible that this has been a contributing factor, especially given that the rapid ramp-up of IP creation output in Singapore since the late 1990s coincided with the establishment of a number of public initiatives targeted specifically at strengthening IP creation and commercialization capabilities. In particular, we can highlight the following recent developments in Singapore that may be relevant:

- The centralization of IP management and commercialization policy among all the public research institutes managed by the Agency for Science, Technology and Research (A*STAR) under one organization, Exploit Technologies (ETPL).
- The establishment of an IP Academy in 2003 to provide executive IP education and to develop thought leadership on IP management in Singapore and Asia.
- The establishment of the IP Office of Singapore (IPOS) in 2001 with a broad mission not only to administer IP laws and to promote IP awareness (as covered by IPD in HKSAR), but also to provide infrastructure support for IP development, including working with the IP business and professional community to identify and develop business opportunities related to IP, especially in terms of making Singapore a regional hub for IP management services and thought leadership (e.g., the convening of the annual Global Forum on IP in 2008). Unlike HKSAR, IPOS implements a local patent examination system, although registration of patent applications submitted to other PCT member jurisdictions is also allowed.
- The active strategic role of the Economic Development Board (EDB) in Singapore (which is the counterpart to Invest Hong Kong) to attract investment and talents to the IP professional services industry.
- The active role of SPRING, the government agency in Singapore in charge of promoting local SMEs, in promoting the development of an early stage angel investment community for high-tech, IP-based start-ups, by providing matching investment funds.

More recently, the Singapore government established the National Research Foundation (NRF), a new R&D funding agency to strategically fund "use-inspired" R&D in targeted new, emerging technologies beyond the current R&D focus of the public research institutes managed by A*STAR (e.g., clean-tech, interactive digital media, biomedical translational research). Besides providing strategically targeted R&D funding, the NRF has also established a comprehensive framework (National Innovation Framework) to facilitate the commercialization of IP generated from the R&D programs. The framework includes funding to improve the IP management and commercialization capabilities of the universities, Small Business Innovation Research (SBIR)-like funding specifically earmarked (as a percentage of R&D funding) for the funded program to explore commercialization feasibility, as well as providing matching funds to grow a number of early stage venture capital funds focusing on high-tech, IP-based start-ups, particularly spin-offs from the local universities.

5. Conclusion

Since 1997, the HKSAR government has certainly done well in terms of several aspects of IP policy: strengthening HKSAR's environment for IPR through the development of its IP legislative framework, improving the effectiveness of its enforcement institutions, and educating the public and raising their awareness of IPR. Nevertheless, based on the observations of this chapter, I believe that a case can be made for a more strategic role of the government of HKSAR in terms of other aspects of IP policy intervention, if the government is indeed committed to driving HKSAR towards a knowledge economy that is based on a more advanced innovation system. In particular, over and above the intensification of public investment in innovation activities in general and R&D activities in particular, the HKSAR government should consider taking a more active policy intervention role in terms of promoting the city economy as a major hub in China and Asia for IP creation, commercialization and transaction. This would include promoting the development of its IP professional services industry and industrial design services industry, and strengthening the IP creation and commercialization capabilities of its universities and public research institutes to serve the needs of both HKSAR and the PRD region.

While this proposed broadening of the government's IP policy role beyond IPR protection would represent a fundamental shift in the government's current laissez-faire "positive non-intervention" approach, I believe that such a broader approach is necessary if HKSAR is serious about developing an innovation-driven economy.

Appendix

Table A.1 Growth of Hong Kong and Singapore patents 1976–2007

	HK assignee patents by HK inventor	Foreign assignee patents by HK inventor	Total patents by HK inventor	Sg assignee patents by Sg inventor	Foreign assignee patents by Sg inventor	Total patents by Sg inventor
			No. of patents			
1976	17	7	24	2	1	3
1977	25	8	33	3	2	5
1978	27	3	30	0	3	3
1979	21	15	36	0	0	0
1980	44	9	53	1	5	6
1981	57	13	70	3	2	5
1982	67	6	73	6	0	6
1983	64	1	65	4	2	6
1984	58	20	78	4	0	4
1985	56	17	73	7	7	14
1986	106	11	117	3	2	5
1987	80	20	100	10	6	16
1988	91	23	114	5	7	12
1989	116	27	143	16	12	28
1990	109	55	164	6	15	21
1991	144	79	223	17	14	31
1992	123	50	173	13	28	41
1993	154	43	197	19	42	61
1994	180	58	238	29	50	79
1995	218	58	276	30	51	81
1996	218	55	273	54	70	124
1997	207	85	292	55	77	132
1998	306	105	411	90	91	181
1999	337	105	442	102	105	207
2000	422	138	560	179	120	299
2001	426	167	593	233	154	387
2002	394	184	578	296	237	533
2003	419	201	620	291	273	564
2004	347	201	548	273	320	593
2005	327	162	489	210	255	465
2006	411	243	654	284	299	583
2007	351	369	720	241	294	535
Total	5922	2538	8460	2486	2544	5030
			Annual growth rate (%)			
1976–86	20.1	4.6	17.2	4.1	7.2	5.2
1986–96	7.5	17.5	8.8	33.5	42.7	37.9
1996–00	18.0	25.9	19.7	34.9	14.4	24.6
2000–07	–2.6	15.1	3.7	4.3	13.7	8.7

Notes:
1: Where a patent is assigned to more than one country, it is allocated according to the country of the first-named company.
2: Patents by Hong Kong (Singapore) inventors include all patents with at least one inventor who is a Hong Kong (Singapore) resident.
3: Unassigned patents are allocated to Hong Kong (Singapore) assignees.
Source: USPTO database and National University of Singapore (NUS) patent database.

Table A.2 Growth of Hong Kong and Singapore utility patents 1976–2006

	HK assignee patents by HK inventor	Foreign assignee patents by HK inventor	Total patents by HK inventor	Sg assignee patents by Sg inventor	Foreign assignee patents by Sg inventor	Total patents by Sg inventor
			No. of patents			
1976	15	7	22	2	1	3
1977	9	4	13	1	2	3
1978	19	3	22	0	3	3
1979	8	10	18	0	0	0
1980	24	7	31	1	4	5
1981	28	8	36	3	2	5
1982	18	5	23	6	0	6
1983	18	1	19	5	1	6
1984	22	8	30	4	0	4
1985	20	12	32	6	7	13
1986	29	8	37	3	1	4
1987	26	13	39	10	5	15
1988	35	14	49	5	4	9
1989	39	17	56	15	7	22
1990	30	30	60	4	12	16
1991	34	23	57	9	12	21
1992	49	21	70	11	27	38
1993	46	24	70	14	41	55
1994	44	31	75	23	48	71
1995	71	36	107	26	45	71
1996	73	36	109	48	57	105
1997	68	32	100	54	67	121
1998	131	61	192	85	70	155
1999	112	68	180	98	99	197
2000	120	88	208	167	107	274
2001	181	92	273	229	144	373
2002	182	98	280	288	217	505
2003	197	99	296	282	241	523
2004	187	110	297	260	284	544
2005	161	78	239	190	242	432
2006	183	131	314	252	264	516
Total	2179	1175	3354	2101	2014	4115
			Annual growth rate (%)			
1976–86	6.8	1.3	5.3	4.1	7.2	5.2
1986–96	9.7	16.2	11.4	33.5	42.7	37.9
1996–00	13.2	25.0	17.5	36.6	17.1	27.1
2000–06	7.3	6.9	7.1	7.1	16.2	11.1

Notes:

1: Where a patent is assigned to more than one country, it is allocated according to the country of the first-named company.

2: Patents by Hong Kong (Singapore) inventors include all patents with at least one inventor who is a Hong Kong (Singapore) resident.

3: Unassigned patents are allocated to Hong Kong (Singapore) assignees.

Source: USPTO database and NUS patent database.

Table A.3 Utility patenting propensity, selected economies, 1985–2005

	Utility Patenting Propensity (Patents per 100,000 population)		
	1985	**1995**	**2005**
Japan	10.59	17.56	24.1
South Korea	0.1	2.62	9.2
Taiwan	0.91	7.83	22.9
Hong Kong	0.59	1.72	3.4
Singapore	0.47	2.0	9.8
China	0	0.01	0.05
India	0	0.01	0.05
USA	16.7	21.2	25.9
Germany	8.73	8.48	11.91
Ireland	0.88	1.83	4.81

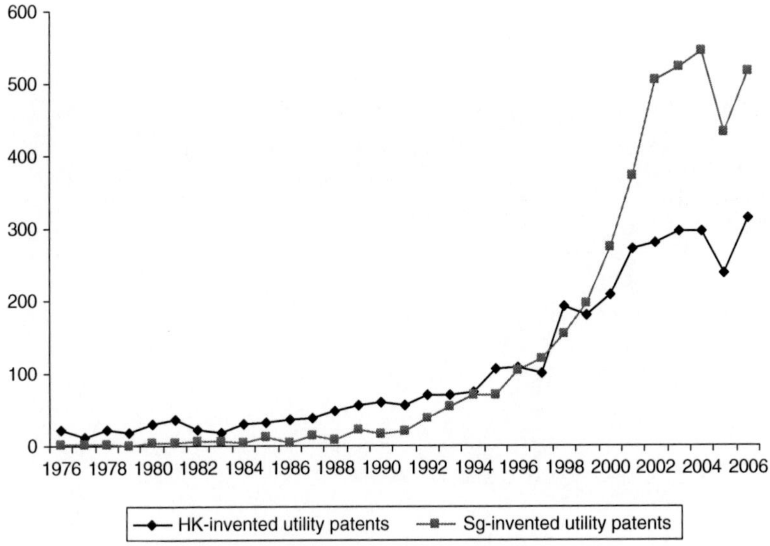

Figure A.1 Growth of Hong Kong-Invented utility patents vs. Singapore-Invented utility patents 1976–2006
Source: USPTO database and NUS patent database.

Table A.4 Trademarks applications and registrations with USPTO

Fiscal Year ending September	Trademark Applications Filed		Trademarks Registered	
	HK	Singapore	HK	Singapore
1990	285	48	82	9
1991	360	58	83	10
1992	484	66	130	17
1993	319	97	175	28
1994	396	172	160	23
1995	456	138	127	33
1996	456	110	168	45
1997	437	203	163	60
1998	478	161	169	49
1999	625	186	146	34
2000	1,097	419	194	44
2001	898	339	267	76
2002	860	283	288	82
2003	794	285	387	95
2004	862	205	391	102
2005	1,130	311	290	100
2006	1,113	355	373	110
2007	1,305	503	424	134
Total (1990–2007)	12,355	3,939	4,017	1,051
	Average Annual Growth (%)			
1990–1995	9.9	23.5	9.1	29.7
1996–2001	14.5	25.2	9.7	11.1
2001–2007	6.4	6.8	8.0	9.9

Table A.5 Comparison of Hong Kong and Singapore patents by patent type 1976–2006

	Hong Kong				Singapore			
	No. of patents				No. of patents			
	Utility	Design	Plant, Tissue	Total	Utility	Design	Plant, Tissue	Total
1976–85	295	318	1	614	56	7	0	63
1986–95	756	1169	2	1927	346	59	0	405
1996–00	920	1326	1	2247	912	95	0	1007
2001–06	2087	2191	7	4285	3253	233	6	3492
	% of patents				% of patents			
1976–85	48.0	51.8	0.2	100.0	88.9	11.1	0.0	100.0
1986–95	39.2	60.7	0.1	100.0	85.4	14.6	0.0	100.0
1996–00	40.9	59.0	0.0	100.0	90.6	9.4	0.0	100.0
2001–06	48.7	51.1	0.2	100.0	93.2	6.7	0.2	100.0

Note: Includes patents by at least one locally resident inventor and patents with the first-named assignee who is listed locally.

Table A.6 Breakdown of patents by Hong Kong and Singapore inventors[1] (Local vs. Foreign Assignee) (1976–2006, Percentage)

	1976–1985	1986–1995	1996–2006	Total	1976–1985	1986–1995	1996–2006	Total
	Hong Kong (% of patents)				**Singapore (% of patents)**			
Local assignee	*81.5*	*75.7*	*69.9*	*72.0*	*57.7*	*39.9*	*50.8*	*50.0*
Private Company	49.5	55.5	48.2	49.9	23.1	21.5	34.8	33.6
University	0.0	0.3	3.7	2.7	0.0	3.5	4.6	4.4
Govt/PRIC	0.0	0.0	0.1	0.1	0.0	0.8	6.0	5.5
Individual/Unassigned	32.0	19.9	17.9	19.3	34.6	14.1	5.4	6.4
Foreign assignee	*18.5*	*24.3*	*30.1*	*28.0*	*42.3*	*60.1*	*49.2*	*50.0*
Private Company	17.8	23.6	29.0	27.0	36.5	58.8	47.5	48.3
University	0.0	0.3	0.3	0.3	3.8	0.5	1.2	1.1
Govt/PRIC	0.0	0.0	0.5	0.3	0.0	0.0	0.4	0.3
Individuals	0.7	0.5	0.3	0.4	1.9	0.8	0.2	0.2
Total	100.0	100.0	100.0	100.0	*100.0*	*100.0*	*100.0*	*100.0*

Notes:
[1] Patents where at least one inventor is a Hong Kong (Singapore) resident.
Unassigned included in individuals.
Based on first-named assignee.
Companies formed to commercialize university technology are counted under universities and PRICs.
Source: USPTO database and NUS patent database.

Table A.7a Average citations received per utility patent by Hong Kong and Singapore inventors 1976–2006

	Hong Kong	Singapore
1976–85	10.9	6.5
1986–95	11.0	12.9
1996–06	4.0	4.2
Overall	5.8	4.9

Note: Computed using citations up to 2006. Because of truncation effect, more recent patents tend to have lower forward citation counts due to having less time to attract forward citations.

Table A.7b Relative Citation Index, 1976–2005

Country of Invention	All Patents				Utility Patents			
	1976–1985	1986–1995	1996–2000	2001–2005	1976–1985	1986–1995	1996–2000	2001–2005
Hong Kong	0.847	0.708	0.745	0.977	1.105	0.943	0.938	1.153
Singapore	0.652	1.116	1.265	1.074	0.641	1.110	1.277	1.187

Table A.7c High Impact Index (top 5 percent most highly cited utility patents within 1-Digit Technology Class), 1976–2005

Country of Invention	Using 1 Digit Technology Class			
	1976–1985	1986–1995	1996–2000	2001–2005
Hong Kong	1.272	0.879	1.204	1.244
Singapore	0.000	0.970	1.616	1.268

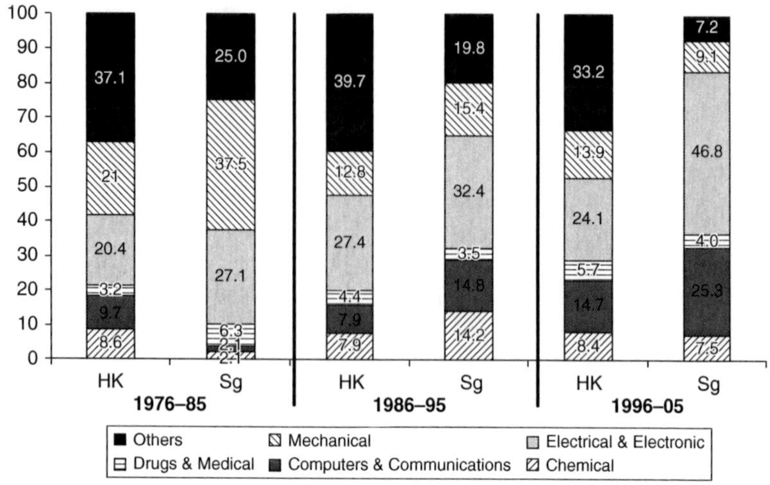

Figure A.2 Comparison of technology class of patents by Hong Kong and Singapore inventors, 1976–2006

Note: Patents include those where at least one inventor is a Hong Kong/Singapore resident.
Source: USPTO database (various years) and NUS patents database.

Table A.8 Herfindahl Index of technological concentration, 1976–2006

	Hong Kong	Singapore
1976–85	0.277	0.281
1986–95	0.203	0.211
1996–00	0.203	0.282
2001–06	0.194	0.312

Notes: Nationality of patent is defined as having at least one inventor resident in the specified nation.

Herfindahl Index computed using classifications at the IPC Section level, with 8 categories in total.

Source: Computed from the USPTO database (various years) and the NUS database of US patents.

Table A.9a Top 20 organizations with Hong Kong patents[1]

No. of Companies		Country	Patent Count				
			1976–85	1986–95	1996–00	2001–06	Cumulative Total as at end 2006
1	John Manufacturing Ltd	Hong Kong	9	170	115	41	335
2	Hong Kong University of Science & Technology	Hong Kong	0	1	27	53	81
3	Vtech Industries, Inc.[2]	Hong Kong	0	17	37	21	75
4	One World Technologies Limited	Hong Kong	0	0	0	70	70
5	Johnson Electric S.A.[3]	Hong Kong	1	49	13	2	65
6	Hayco Manufacturing Limited	Hong Kong	0	0	2	61	63
7	Astec International Limited	US	0	34	16	12	62
8	Choon Nang Electrical Appliance	Hong Kong	0	3	16	36	55
9	The Brinkman Corporation	US	0	8	2	42	52
10	Hong Kong Polytechnic University	Hong Kong	0	0	5	45	50
11	World Wide Stationary Manufacturing Co. Ltd	Hong Kong	0	5	21	22	48
12	SAE Magnetics (Hong Kong)	Hong Kong	0	0	1	45	46
13	Gold Coral International Ltd	Hong Kong	0	0	0	45	45
13	Solar Wide Industrial Limited	Hong Kong	0	17	17	11	45
15	Rosalco Inc.	US	0	44	0	0	44
15	Koninklijke Philips Electronics N.V.[4]	Netherlands	7	8	8	21	44
17	STD Electronic International[5]	Hong Kong	0	38	4	1	43
18	Motorola Inc.	US	0	11	22	9	42
19	Goodway Electrical Company Ltd	Hong Kong	1	7	8	25	41
20	Alfa Technology Ltd	Hong Kong	0	0	30	10	40
20	Timex Corp[6]	US	3	1	20	16	40

[1] Patents where at least one inventor is a Hong Kong resident. The first assignee company is used to count patents which are assigned to more than one company.
[2] Includes Vtech Communications Ltd, Vtech Electronics Limited, VTech Telecommunications Limited, Vtechsoft Holdings Limited.
[3] Includes Johnson Electric Engineering Ltd, Johnson Electric Industrial Manufactory.
[4] Includes North American Philips Corp., US Philips Corp.
[5] Includes STD Manufacturing Ltd, STD Plastic Industrial Ltd.
[6] Includes Timex Group B.V.
Source: NUS patents database.

Table A.9b Top 20 organizations with Singapore patents[1]

No. of Companies		Country	Patent Count			
			1986–1995	1996–2000	2001–2006	Cumulative Total as at end 2006
1	Chartered Semiconductor Manufacturing	Singapore	14	190	581	785
2	Hewlett-Packard Company	United States	23	52	141	216
3	Seagate Technology	United States	0	15	200	215
4	National University of Singapore	Singapore	12	35	115	162
5	Micron Technology Inc.	United States	0	0	135	135
6	Motorola Inc.	United States	24	47	34	105
7	Texas Instruments	United States	18	42	37	97
8	Koninklijke Philips Electronics N.V.[2]	Netherlands	10	15	63	89
9	Institute of Microelectronics	Singapore	1	18	62	81
10	ST Assembly Test Services	*Singapore*	1	*2*	*78*	81
11	Matsushita Electric Industrial Co.	Japan	3	24	49	76
12	Agency for Science, Technology and Research	Singapore	0	0	57	57
13	ST Microelectronics	Italy/France	2	17	38	57
14	*Tri-tech Microelectronics*[3]	United States	3	49	4	56
15	Creative Technology	Singapore	0	9	46	55
16	Advanced Micro Devices	United States	0	9	43	52
17	Thomson SA	France	15	10	20	45
17	ASM International NV[4]	Netherlands	0	0	41	41
19	Infineon Technologies	Germany	0	0	41	41
20	Molex Incorporated	United States	26	7	5	38

[1] Patents where at least one inventor is a Singapore resident. The first assignee company is used to count patents which are assigned to more than one company.
[2] Includes US Philips Corp.
[3] A company called Tri-tech Microelectronics was granted a total of 56 patents before filing for bankruptcy and entering liquidation in 1999.
[4] Includes ASM Technology Singapore.
Source: USPTO Database (various years).

Note

1. For contextual background information on Hong Kong's economic and innovation systems, please refer to: Baark and Naubahar (2006); GEM (2007); Hong Kong Census and Statistics Department (2007a, 2007b); HKUST Office of Contract & Grant Administration (2008); HKUST Technology Transfer Center (2008); ITC website; Hong Kong Science & Technology Parks website; Digital Asset Management Project website.

References

Baark, Erik and Naubahar Sharif (2006) "Hong Kong's innovation system in transition" in *Asian Innovation Systems in Transition*, edited by Bengt-Åke Lundvall, Patarapong Intarakumnerd, and Jan Vang. New York: Palgrave Macmillan.

Berger, S. and R. K. Lester (eds) (1997) *Made by Hong Kong*. Hong Kong: Oxford University Press.

Census and Statistics Department, HKSAR (2008) *Report on 2006 Annual Survey of Innovation Activities in the Business Sector* (Hong Kong).

Database of the USPTO: http://patft.uspto.gov/.

Digital Asset Management Project website: http://www.cyberport.hk/cyberport/en/home/facilities_n_services/irc/damp/. Accessed on 1 January 2008.

Enright, M. J., E. E. Scott and D. Dodwell (1997) *The Hong Kong Advantage*. New York: Oxford University Press.

GEM (2007) *Global Entrepreneurship Monitor: Hong Kong 2007*. Hong Kong: Center for Entrepreneurship, The Chinese University of Hong Kong.

Hong Kong Census and Statistics Department (2007a) *Hong Kong Monthly Digest of Statistics: Feature Article – Statistics on R&D of Hong Kong, 1998–2002 & 2001–2005*. C&SD, HKSAR.

Hong Kong Census and Statistics Department (2007b) *Hong Kong as a Knowledge-based Economy: A Statistical Perspective, 2005 Edition*, C&SD, HKSAR.

Hong Kong University of Science and Technology (HKUST) Office of Contract & Grant Administration (2008) *Funding Statistics: Research Funding, Hong Kong University of Science & Technology*. From http://www.ogca.ust.hk. Accessed on 1 January 2008 (protected?).

Hong Kong Science & Technology Parks (2008). http://www.hkstp.org/HKSTPC/en_html/en_index.jsp. Accessed on 1 January 2008.

Hong Kong University of Science & Technology (HKUST) Technology Transfer Center (2008) Patent Statistics. From http://www.ttc.ust.hk/patents3/index5.htm. (2000–2001 & 2005–2006 figures) Accessed on 1 January 2008 (database).

Idris, K. (2003) *Intellectual Property: A Power Tool for Economic Growth*, WIPO, 2nd edition (June 2003).

Innovation and Technology Commission: http://www.itc.gov.hk/en/welcome.htm. Accessed 1 January 2008.

Intellectual Property (IP) Department (2004) *Intellectual Property in Hong Kong*. IPD, Hong Kong. From http://www.ipd.gov.hk/eng/pub_press/publications/iphk_e.pdf.

Intellectual Property (IP) Department: http://www.ipd.gov.hk/eng/home.htm. Accessed on 1 January 2008.

Jaffe, A. B. and M. Trajtenberg (2002) *Patents, Citations, and Innovations: A Window on the Knowledge Economy*. MIT Press, Cambridge, MA.

Japan Strategic Council on Intellectual Property(JSCIP) (2002) "Intellectual Property Policy Outline", http://www.kantei.go.jp/foreign/policy/titeki/kettei/020703taikou_e.html. Accessed on 1 January 2008.

Jinan University(2004) *Research Report on Intellectual Property Professional Support Services in Guangdong and Hong Kong (abstract)*, Intellectual Property School, Jinan University, Guangzhou, China, December 2004.

Lam, C. (2006) *IP Value 2006: China & Hong Kong. Recent Developments in Intellectual Property*. http://www.dorsey.com/files/tbl_s21Publications/PDFUpload141/1471/Lam_IPValue2006.pdf. Accessed on 1 January 2008.

Leung, M.C. (2008) *Origins and Dynamics of University Spin-offs and Start-up Enterprises in Hong Kong*. DBA thesis, Macquarie Graduate School of Management, Macquarie University, Australia.

Mowery, D. C. (2008) "University-industry collaboration and technology transfer in Hong Kong and knowledge-based economic growth". Draft.

Park, W. G. and S. Wagh (2002) "Index of patent rights", chapter 2 in *Economic Freedom of the World: 2002 Annual Report*, Fraser Institute.

Park, W. G. (2001) "Intellectual property and patent regimes", chapter 4 in *Economic Freedom of the World: 2001 Annual Report*, Fraser Institute.

Sharif, N. and E. Baark (2009) "The Hong Kong University of Science and Technology: A Case Study in Entrepreneurial-University-Led Knowledge Based Economic Development". In *The Role and Impact of Universities in a National Innovation System*. Vol. 1 of *A SPF Project Report on Initiatives in Comprehensive Understanding of Civilizational Issues: A New Era of Science and Bioethics*. Singapore: NUS Entrepreneurship Centre.

UNIDO(2006) *The Role of Intellectual Property Rights in Technology Transfer and Economic Growth: Theory & Evidence*, UNIDO working chapter, 2006.

World Economic Forum (various years), *Global Competitiveness Report*. Geneva : World Economic Forum.

Wright, M., A. Lockett, B. Clarysse, and M. Binks (2006) "University spin-out companies and venture capital", *Research Policy*, 35(4): 481–501.

3
Human Resources: Hong Kong's Challenges and Opportunities

David M. Hart and Fangmeng Tian
George Mason University

1. Introduction: The moving target

Hong Kong's recent economic history, viewed from a distance, is a success story. Most obviously, real per capita income has risen sixfold in less than 40 years (USDA, 2008). The region has maintained its economic momentum despite the enormous structural shift entailed by the rapid development since 1978 of Mainland China as a whole and of the Pearl River Delta area in particular (Berger and Lester, 1997; Enright et al., 2005). The economy now relies predominantly on services, notably the four pillar industries identified by the Special Administrative Region (SAR) administration: finance, logistics, tourism, and information services (Tsang, 2007).

In the future, Hong Kong will continue to face a complex and dynamic strategic environment that poses hard choices as well as enormous opportunities. There is no time to bask in the warm glow of past success. Mainland China is moving aggressively beyond its manufacturing-heavy strategy, seeking to upgrade its economy by using and creating technology, science, and innovation ("Guidelines," 2006). In doing so, it is following in the footsteps of the rest of East Asia, which is quickly converging with the West in this regard (Hu and Mathews, 2005). Global industry is in a phase of massive restructuring as well, in response to the growth of emerging markets, the opportunities created by new technologies, and now the financial crisis and consequent recession.

Some of the most important choices that Hong Kong must make in order to sustain its economic success involve human resource development. If it is true that knowledge is the most valuable asset in the twenty-first-century economy, then educated and creative people are its most essential inputs. The Hong Kong population, though, is aging rapidly, and its fertility rate is among the lowest in the world. Hong Kong's institutions of higher education and research are characterized by entrepreneurship and excellence, but also by inertia. Immigration from the Chinese Mainland could be a huge source of skills and knowledge, but the flow must be managed carefully lest public opinion turn sour on it. The alternative of attracting talent from the rest of the world is complicated not only by Hong Kong's unique geopolitical situation, but also by an intensifying competition for highly skilled migrants among the leading economies.

This chapter analyzes Hong Kong's talent pool. We describe stocks and flows of several key human resource indicators, both general and specialized to science and technology. We find that Hong Kong's working population is acquiring the skills and knowledge required to support innovation fairly rapidly. We do not draw a conclusion as to whether the pace is fast enough, given the region's changing context. However, we believe this question to be worthy of further serious consideration and, in light of that, we explore potential policy options for accelerating the pace, should that be desired. Before we get to these main tasks, however, we turn our attention first to the role of human capital in economic development, to provide a context for the numbers and policy options that follow.

2. Human capital and growth: Four cases

In the early 1960s, economists began to recognize formally what historians and other social scientists had long taken for granted: that skills and knowledge embodied in people contribute powerfully to economic growth. In the late 1980s, these insights were developed into full-fledged theories of "endogenous growth" by Lucas (1988) and Romer (1990), among others. Quantitative research grounded in these theories has convincingly demonstrated a correlation between measures of skills and knowledge on the one hand and income per capita at the national, regional, and individual levels on the other (Abel and Gabe, 2008 and references therein).

Although the correlation is strong, the causal relationship between human capital[1] and economic development is anything but simple. In our view, the causal arrows may plausibly go in both directions. A rich supply of human capital can create opportunities for capital investment, enterprise formation, and innovation, boosting economic growth. But it is also the case that rising incomes can provide incentives for individuals and governments to invest in education, skill acquisition, and knowledge creation. Ideally, a virtuous, self-reinforcing circle is established between the economic growth process and the human capital formation process.

These reciprocal causal relationships may hold true, we would suggest, for (1) general human capital and overall economic growth and (2) specialized human capital in science and technology (S&T) fields and industry-level growth in innovation-oriented industrial sectors. However, in both directions of causation, and for both general and industry-level growth, other variables may confound the relationship, especially in a small and open economy like Hong Kong's. In the remainder of this section, we briefly lay out four cases that this line of thought yields.

Let us take as the first case exogenous growth in general human capital, in other words, a broad "supply push." Would an economy expand in response? The answer that the literature gives is "yes," but only over the long run and at the highest level of aggregation (i.e., at the global level in a globalized economy). The key response factors are technology and management (Acemoglu, 2002). As skills and knowledge become more widely available and less costly, new systems of production that take advantage of them are invented. The information revolution of the past few decades, for instance, has depended on the highly educated workforce that became available in that period; effective IT-using organizations demand high skill levels, rather than simply automating tasks in a fashion better suited to the unskilled workforce of the industrial era.

In the short term and for a small economy like Hong Kong's, though, the adjustment envisioned in this case could be very slow or even fail to materialize. If the supply of general human capital is expanding simultaneously in many countries (or even in a single very large country like Mainland China), the suppliers of other factors of production, especially capital and entrepreneurial know-how, may have a choice of investment sites. Even local investors, in an age of footloose capital, may seek the best deal they can find anywhere in the world. Some attention to labor market demand would therefore be a useful check on a general human capital "supply-push" policy. Weak labor market demand does not necessarily indicate that supply should be constrained. It might be an indication that additional policies are required in order to attract or develop other factors of production that are complementary to the newly created human capital.

Now let us consider causation in the opposite direction: would a sustained strong demand for knowledge and skills generated by a thriving modern economy stimulate human capital formation? Here, too, the answer, in general, is "yes" over the long term. As workers recognize that they can make more money over their lifetimes if they make an upfront investment in knowledge and skill acquisition, they should choose to do so – if they can.

But sometimes they cannot. Without public subsidies, they may lack the funds to make such an investment. Or, the educational system may not be flexible enough to accommodate the growing demand. In addition, older workers may be reluctant to invest in education, which might put at risk the value of their accumulated experience. Immigration is another possible source of human capital, but it depends on at least two uncertain factors, the perception of opportunity by potential immigrants and openness to immigration by the local population. Institutional bottlenecks and information barriers, then, are clear dangers to a non-interventionist human resource development policy that relies primarily on labor market demand.

If we turn to the cases that relate specialized human capital to growth in innovation-oriented industries, we find that additional conditions must be taken into account, beyond those present in the first two cases. Our third case assumes an expansion in the supply of technical experts trained to work in a particular industry. The key concept we need to add in this case is industrial clustering. Establishments in the same line of business tend to be located near one another, and the resulting geographical "agglomerations" tend to endure over time. Clustering benefits both the workers who possess specialized human capital and the employers who hire them. Workers prefer to live where they will have a wide range of career options; firms like to have many talented job candidates to choose from. Another reason for clustering is localized knowledge spillovers. The hottest ideas are passed by word of mouth through social networks; geographical proximity allows workers and firms to take advantage of these.

One danger, then, of a specialized human capital "supply-push" policy is emigration by the newly trained specialists to established clusters abroad. This argument is compatible with a policy that concentrates on human resource development for industries in which a location is already specialized. Alternatively, it might be part of a broader case for a sustained, multi-pronged investment that extends beyond human resources to other factors of production in the hope that a new cluster might be established.

The fourth and final case contemplates whether specialized human capital would be formed in response to the expansion of industrial clusters that use the relevant skills and knowledge. Although the supply bottlenecks and information barriers that we discussed in the more general case above would likely be present in this specialized case as well, they may not be as severe. The very prominence of such clusters in the region's economic profile should draw the attention of people, both domestically and overseas, who have or may be able to acquire sector-specific knowledge and skills. The growing clusters' financial and political clout should also help to induce change in educational institutions, and it may also facilitate liberalization of targeted immigration policies.

Looking forward to the empirical section of this chapter, these four theoretical discussions suggest the importance of attending to both supply and demand indicators in assessing Hong Kong's human resource situation. Foreign competition, demographic trends, institutional rigidities, and misperceptions among firms and workers are some of the obstacles that may impede balanced growth of capabilities and opportunities. Our argument to this point provides some warrant for policy intervention, because of these imperfections in the labor market. But policymakers must also take into account the complexity that we have sketched if their efforts are to hit the mark.

It would be remiss of us if we concluded this section without stressing that there are very good reasons, besides economic growth, for individuals and societies to invest in people. Education, skill acquisition, and knowledge creation are virtuous activities in their own right. They help people to fulfill their creative potential, to become more autonomous, and to experience a better quality of life. A view of workers that reduces them to mere factors of production would be cramped indeed. In what follows, then, we will lean toward more such investment in human resources, rather than less, whenever a reasonable case can be made.

3. Human resource indicators: Stocks, flows, and comparisons

This section provides the empirical basis for understanding Hong Kong's human resources. Our conclusion on this point is unequivocal: whether viewed through the lens of supply or demand, the evidence shows that Hong Kong *is moving* toward a better educated workforce that holds higher-status jobs. Whether it is doing so quickly enough, given the strategic context, is more difficult to say. What public policymakers ought to do to hasten the process, should they so choose, is equally challenging. This section provides some insights into these matters, and they are taken up more directly and more fully in the final section of this chapter.

Human capital is intrinsically difficult to measure. It encompasses skills and knowledge, both formal and tacit. There are two ways to acquire it: education and experience. We therefore use two indicators to assess Hong Kong's human resources, looking first at educational attainment and then at occupational status. For each of these indicators, we first consider all fields of activity and then S&T fields in particular. This multiple-indicator approach allows us to paint a fuller picture than any single indicator would. However, we do not seek here to assess the quality of education or experience, which would enrich the portrait further. It is important to stress that the indicators we employ are only imperfect proxies for human capabilities, a complex and elusive real variable.

3.1. Educational attainment: Undergraduate degrees, all fields

Although there is no firm cut-off for entry into knowledge-based work (indeed, accomplished teenage computer hackers are far from unheard of), a university education is a reasonable prerequisite for most such work. Undergraduate education provides advanced competencies in specific disciplines. The high level of knowledge and the degree of specialization are both important in fostering autonomous judgment and creativity. The higher education system sorts students in addition to training them and signals employers about their potential economic contributions.

The number of Hong Kong workers who hold an undergraduate degree, both male and female in roughly equal proportions, has risen rapidly in recent years.[2] As Table 3.1 shows, the total went up by about 70 percent between 1996 and 2006, an annualized growth rate of 5.5 percent. Since the working population of the SAR has grown by

Table 3.1 Undergraduate degree holders as a share of working population

Education Attainment	1996	2001	2006
Degree holders in working population	368,000 (est.)	494,560	627,140
Working population	3,043,698	3,252,706	3,365,736
Share	12.1%	15.2%	18.6%

Source: Hong Kong By-Census, 1996–2006.

only about 1 percent per year during this period, the proportion of this population holding an undergraduate degree has grown substantially, from an estimated 12.1 to 18.6 percent.[3]

The sources of this growth, a positive net flow into the working population of some 26,000 undergraduate degree holders per year, are complex. Demographic replacement accounts for some of it. Younger Hong Kong residents are more likely to hold undergraduate degrees than their parents or grandparents. Well under 10 percent of those aged 60–64 years who were in the labor force in 1996, for instance, held such degrees, compared with almost 30 percent of their counterparts two generations later (those aged 25–29 years who were in the labor force in 2006). However, we find rising levels of educational attainment within many such cohorts as well. For instance, a higher percentage of those aged 50–54 who were in the labor force in 2006 held undergraduate degrees than of those aged 40–44 who were in the labor force in 1996. The same pattern holds for all the younger cohorts in these years as well.

The University Grants Committee (UGC)-funded programs of Hong Kong's most prestigious higher education institutions have made a steady contribution to educating the traditional university-age cohort, producing about 15,000 undergraduate degrees per year since 1997. The growth rate has been less than 1 percent per year.[4] Undergraduate degree programs not funded by UGC have grown more rapidly. They now amount to some 5,000 annually, up from about 2,000 10 years ago.[5] The institutions granting these degrees include the self-financing arms of UGC-funded institutions and five others that receive no UGC support.

The rest of the inflow is comprised of (1) Hong Kong people obtaining degrees abroad and later returning home, and (2) immigrants and temporary residents. The latter group is easier to estimate. The General Employment Policy (GEP) for in-migration for the purpose of high-skill work has grown by about 50 percent in recent years, from about 15,000 annually in the late 1990s to almost 22,000 in 2006.[6] About three-quarters of these workers are in occupations that are highly likely to require a university education. High-skill immigration from Mainland China is handled through the Admission Scheme for Mainland Talents and Professionals (ASMTP), which was initiated in 2003. In 2006, about 5,000 Mainlanders were admitted under this program, virtually all of whom held an undergraduate degree.[7]

We can calculate only indirectly and roughly the number of those who leave Hong Kong to get a university degree and then later return. Relying primarily on census data, we estimate that about 6,500 such individuals returned each year between 2001 and 2006. This figure is somewhat lower than that of the previous 5-year period (1996–2001), which may be accounted for by the uncertainty that attended the 1997 transition to Chinese sovereignty and its positive resolution, which encouraged return in the post-1997 period.

We do not have direct evidence about departures from Hong Kong of university degree holders, whether Hong Kong- or Mainland-born or otherwise. Our estimates of the inflows to the resident population for 2006 add up to about 48,000 (20,000 from local universities, 16,500 under the GEP, 5,000 under ASMTP, and 6,500 returnees with non-local degrees) (see Figure 3.1). If we assume, conservatively, that 80 percent (38,000) of this inflow to the resident population joins the working population, we can estimate, very imprecisely, that about 12,000 degree holders depart each year, based on an annual growth in the working population of about 26,000 undergraduate degree holders.[8]

We can conclude that Hong Kong has accumulated general human resources, as measured by its university-educated population, steadily and rapidly over the past decade. The growth has been made possible by the SAR's liberal immigration policy, the willingness of its residents to travel (and pay) for higher education abroad, and the emergence of self-financing degree-granting institutions.

Human capital theory argues that investments in education are recouped through higher earning power over the entire life cycle. We do not have long-term data to test this proposition. However, we can

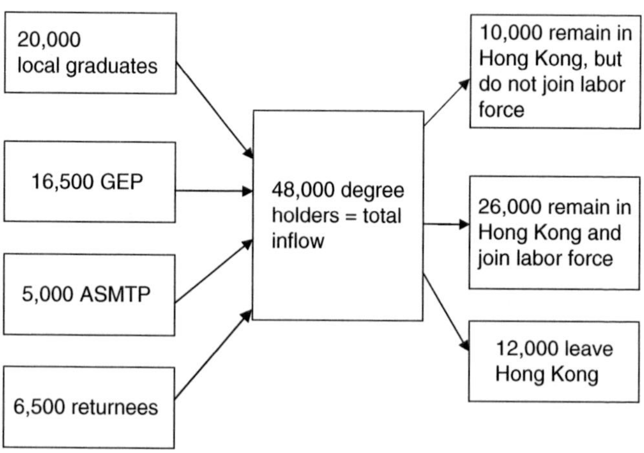

Figure 3.1 Estimated flows of undergraduate degree holders in 2006
Source: Authors.

look at short-term indicators of demand that may provide insight into whether Hong Kong people will continue to make investments in their own human capital.

These data generally suggest that they will.[9] Employment of undergraduate degree holders, for example, has grown steadily each year in recent years. Unemployment amongst this group has trended down and, as of the end of 2007, was about 2 percent. Cross-sectional income data from the census show that the gap between those who have attended university degree courses and those who have not widened considerably between 1996 and 2006.[10] Time-series income data, on the other hand, inject a note of caution on this issue. During the late 1990s' boom, the earned incomes of the highly educated went up, but these gains were largely given up during the recession of the early 2000s. Indeed, some in this group, especially those earning the highest incomes before the recession, may still be worse off today than they were in 2001.

Two simple extrapolations provide a sense of the impact if the 10-year trend in growth of undergraduate degree holders were to continue into the future. If the growth rate of 5.5 percent per year was maintained, the number of Hong Kong workers holding an undergraduate degree would double in about 13 years. A more conservative approach, using a simple linear projection, predicts a rise of about 50 percent in this population over this period. Of course, many other factors, from Hong Kong's low birth rate and the potential for expanded immigration from the Mainland to the evolving structure of the Hong Kong economy, would need to be considered to assess whether basic "business-as-usual" scenarios like these would be realized.[11]

International comparisons provide another lens through which to view Hong Kong's position. To be sure, such comparisons are rough, due to differences among statistical authorities in their classification schemes and data collection methods. Still, allowing a substantial cushion for error, these comparisons suggest that Hong Kong would need to sustain, and perhaps accelerate, its human capital accumulation process if it were to seek to match other "global cities" and other small, rapidly growing "tiger" economies. For instance, among large cities in the Asia-Pacific region and global financial centers, Hong Kong's share of the tertiary-educated population lies in the same tier (10–15 percent) as Beijing, Shanghai, and Singapore. It lags significantly behind Vancouver and Tokyo (20–25 percent) and far behind Taipei, New York, San Francisco, and London (30–35 percent). Hong Kong also seems to be a few years behind the economies of Taiwan and Ireland in this respect. About 30 percent of those economies' adult populations have some tertiary education, although the share of undergraduate degree holders is smaller than that. Investors may be interested in the size of the talent pool as well as its composition. Viewed this way, Hong Kong is comparable to Taipei and more attractive than Singapore.[12]

As we will discuss in more detail below, higher education in Hong Kong confronts a tension between positive trends, even higher aspirations, and institutional and political constraints. The SAR's recent accomplishments are impressive. But, given the sources of growth in the recent period, pushing the pace – and perhaps even maintaining it – may force policymakers to confront difficult trade-offs.

3.2. Educational attainment: Tertiary degrees, science and technology fields

Science and technology are important resources that innovation-oriented industries must draw upon. Specialists in S&T fields are required to generate discoveries, inventions, and new products and processes. To an important degree, these specialists are also necessary to use effectively S&T generated elsewhere. Advanced training in S&T fields creates a pool of talent that can participate in these tasks.

The reader should note that there is a substantial difference between those who receive training in S&T fields and those employed in S&T-related occupations. Many who hold S&T degrees do not work in S&T-related occupations; some who work in S&T occupations do not have educational backgrounds in S&T fields. Occupational data are explored below.

The number of Hong Kong workers with undergraduate degrees in S&T fields[13] grew at roughly the same rate as that for the broader tertiary-educated working population. We estimate that the total rose by about 62 percent between 1996 and 2006, an annualized growth rate of about 5 percent (compared with 5.5 percent for degrees in all fields).[14] As Table 3.2 shows, the share of S&T undergraduate degree holders in the working population rose over this period from 4.9 to

Table 3.2 S&T degree holders and share in working population

Education Attainment	1996	2001	2006
S&T degree holders in working population	148,500 (est.)	198,720	241,114
Working population	3,043,698	3,252,706	3,365,736
Share (S&T/working population)	4.9%	6.1%	7.2%

Source: Hong Kong By-Census, 1996–2006.

7.2 percent. Although the number of females within this population grew at a slightly faster pace than the number of males, the female share remained below a third in 2006.

Some in Hong Kong have expressed concern that recent graduates are less inclined to select S&T fields than in the past. Census data do not confirm these fears, however. If anything, younger people who have received an undergraduate degree are slightly more likely to hold it in an S&T field than their older counterparts.

The annual net inflow to the population in Hong Kong holding an S&T undergraduate degree averaged about 9,300 over this decade.[15] The annual contribution of UGC-funded programs is not far from that. It rose from about 8,200 in 1997 to 9,600 in 2004, before declining slightly to about 8,600 in 2007. We do not have specific data on the contributions of non-UGC-funded institutions to this pool, but our impression is that the contributions are smaller in these fields than in others, such as business. The 2006 census shows that there were about 39,000 holders of S&T degrees from local institutions in the 25–29 age group, which compares with over 46,500 graduates with S&T degrees from UGC-funded institutions in the 1999–2003 period.

The difference in these two figures implies that quite a few graduates from UGC-funded institutions in S&T fields did not join the working population or left Hong Kong between the time of graduation and the 2006 census.[16] Data on in-migration of S&T undergraduate degree holders into the resident population seem to confirm the point. About 2,500 Hong Kong-born S&T undergraduate degree holders returned annually between 2001 and 2006. About 3,000 foreign-born S&T undergraduate degree holders per year, who had not lived in Hong Kong in 2001, were living there in 2006.[17]

Labor market data are somewhat less encouraging for graduates in S&T fields than for undergraduate degree holders in general. The share of S&T degree holders among all degree holders in the working population declined slightly between 1996 and 2006, from 40.4 to 38.4 percent, which means that their employment growth was less rapid. Salaries, at least since 2000, have not kept up, according to UGC and census figures. Starting salaries for graduates of UGC-funded undergraduate programs across all S&T fields peaked in that year. As of 2006, they still stood about 10 percent below the peak and lagged behind other fields. Of course, these short-term data are imperfect proxies for the perceptions of opportunities of the entire life cycle that human capital theory suggests drive students' decisions about their major field. Perhaps more importantly, these data aggregate across a wide range of fields. Faculty members and employers interviewed for this project suggest that there may be shortages in specific subdisciplines of engineering.

Since the trends are the same among S&T degree holders as among the broader tertiary-educated population, our projections are similar. The doubling time of income based on growth rates, for instance, is 14 years for S&T degree holders (compared with 13 years for all degree holders). International comparisons for this indicator are limited to only two other cities and must be taken only as suggestive, due to definitional differences and varying dates of data collection. Measured as a share of the population, Hong Kong's S&T talent pool is about the same as Singapore's, but it lags far behind Vancouver's. However, in absolute size – which may be important to investors concerned about recruiting an S&T workforce – Hong Kong's pool of S&T undergraduate degree holders is significantly larger than Singapore's.

Our analysis of the limited data on S&T degree recipients varies slightly from our analysis of higher education as a whole. Both supply and demand in S&T fields have lagged slightly behind all fields combined, with demand a little weaker than supply. These findings suggest that any concerted effort to expand S&T enrollments among undergraduates should be undertaken carefully. The occupational pay-off from such an educational investment might be limited unless S&T graduates' salaries grow faster as well, since those trained in these fields may choose other fields upon graduation. As one interviewee put it, many such graduates already go into financial services and "never look back."

3.3. Occupational status: Professionals and managers

While tertiary education may prepare people to engage in knowledge-based work, they may choose not to undertake such work or be unable to find it. More importantly, some of those whose occupations involve considerable autonomous judgment and creativity have gained these skills through their work and life experiences, rather than through formal schooling. Thus, although there is overlap between them, occupational status data provide a perspective complementary to that of educational attainment data.

We focus in this subsection on high occupational status (HOS), which includes all jobs classified as managers and administrators, professionals, and associate professionals. About 1.1 million Hong Kong residents held such positions in 2006, accounting for about a third of all workers (see Table 3.3). The growth rate of these occupations over the past 10 years was about 2.2 percent per year, somewhat faster than that of the labor

Table 3.3 High occupational status (HOS)

Year	Total HOS	Total employment	Share
1996	890,046	3,043,698	29.2%
2001	1,028,133	3,252,706	31.6%
2006	1,109,635	3,365,736	33.0%

Source: Hong Kong By-Census, 1996–2006.

force as a whole, but much slower than that of the tertiary-educated population.

Despite the large difference in the growth rates of the two indicators, the HOS population remains much bigger than that of the tertiary-educated. About 43 percent of Hong Kong residents holding HOS positions in 2006 had received undergraduate degrees, up from an estimated 33 percent in 1996. The higher educational attainment in the HOS workforce reflects the replacement of older, less educated workers by their better educated children and grandchildren.[18] On the other hand, of all undergraduate degree holders in the working population in 2006, about 77 percent held HOS positions, a decline of about 5 percentage points from 5 years earlier. In addition, the bulk of the growth in HOS positions over the past decade – almost 80 percent of it – has been among associate professionals; the number of managers and administrators, as measured by the census, actually shrank during that period.[19]

Growth in the HOS population is not even over time. In the year of its most rapid expansion (according to the General Household Survey (GHS)), it added about 85,000 Hong Kong residents, while in lean times, it shrank. Although the average net inflow between 1996 and 2006 was 20,000–25,000 per year, the short-term fluctuations make it difficult to account for the flows into and out of HOS positions with much precision. However, we can say with some confidence that the cumulative growth is almost entirely among the Hong Kong-born. The number of foreign-born and Mainland-born residents in HOS positions declined slightly between 1996 and 2006. Hong Kong-born residents in such positions numbered about 40 percent more in 2006 than in 1996.

The in-migration pattern of holders of HOS positions has been stable. About 6 percent of this group reported living outside Hong Kong 5 years earlier in both 2001 and 2006. Among HOS position-holders who held graduate degrees, there was a very large growth in the number holding non-local and distance-learning degrees, relative to those holding local degrees, between 2001 and 2006. This may be interpreted as a promising sign that Hong Kong is effectively attracting back students who go to university abroad, or as an indicator that the talent pool for local universities to draw upon in recruiting students (who would otherwise go abroad) is deeper than some think.

The annual fluctuations in HOS positions point to an important difference between the educational and occupational approaches to assessing the Hong Kong talent pool. A degree takes a long time to complete, while a new job title can be gained overnight. The HOS population is therefore much more elastic than the population holding a graduate degree, and it responds more quickly and directly to economic incentives.

Our findings about economic incentives are interesting. Salary indices for HOS positions have risen fairly steadily over the past decade or so, particularly if one looks at career paths rather than at starting salaries.[20] However, a closer look within this population (using census data) shows that the incomes of managers and administrators rose more than those of professionals and associate professionals, and sustained their gains through the recession of the early 2000s. This difference may indicate a shortage of managers and administrators, especially in light of the fact that the size of this group shrank slightly during these years. Unemployment is also extraordinarily low (1 percent or below) among managers and administrators, and among professionals.

At its current rate of growth, the doubling time for the HOS population is about 30 years, about twice as fast as the working population as a whole. More rapid growth would depend on both the overall pace of economic growth and the structure and composition of Hong Kong enterprises. Other things being equal, a more rapidly expanding pool of highly educated workers might also enable such an acceleration, although the correlation between HOS and educational attainment is imperfect.

International comparisons using occupational data are somewhat more favorable to Hong Kong than those using educational data. Relative to other Asia-Pacific and "global" cities, the share of Hong Kong's labor force that holds HOS positions is quite a bit larger than Beijing's and on a par with Singapore's. It is about half the size of London, New York, and San Francisco. It is also roughly the same share as those of Taiwan, Ireland, and Israel, economies which have become centers of high-technology industry in recent years.

Data on high occupational status show that Hong Kong employers have been able to fill most HOS positions, even though the pool of degree-holding workers is not that large relative to their needs. Although the overall trend for HOS employment is positive, employers reduce the HOS headcount with alacrity when the Hong Kong economy slows. This pattern seems to characterize associate professionals especially, who comprise most of the HOS group. We would expect it to continue, since the degree-holding population continues to grow quite rapidly. Managers and administrators, particularly at the top level, have been harder to find, and their lack of availability may be a constraint on growth. This finding may imply that any policy program should emphasize practical experience and managerial training as well as general and technical education.

3.4. Occupational status: S&T-related occupations

The S&T workforce can be defined in many ways. The US National Science Board provides estimates for the US that vary by a factor of four (from 5 to 20 million), based on differing combinations of educational attainment and occupational status (US National Science Board, 2008). Far more Americans hold an S&T degree as their highest degree than work in S&T jobs. To the best of our knowledge, the Hong Kong government does not compile estimates of the S&T working population that are comparable to those of the US.

However, the SAR does gather data on two occupational groups that are important components of an economy capable of supporting innovation. Research and development (R&D) personnel perform the functions of discovering new scientific knowledge, improving production processes, and generating new products. It is important to note that other workers who are *not* accounted for in R&D may also perform these functions, especially in the service sector. The figures reported in this section should be seen as indicative, rather than definitive. A similar caveat applies to the high-skill IT workforce, which is the second indicator that we focus on here. Many workers, other than those whose job title place them in this category, may modify or improve IT systems.[21]

Hong Kong's R&D workforce has been growing very rapidly and steadily, especially since 2000. Between 1998 and 2006, it grew by 154 percent, a growth rate of about 12 percent per year, to about 23,000.[22] The R&D workforce employed by Hong Kong businesses (as opposed to government or higher education) accounted for the lion's share of this growth. From less than half the size of the academic R&D workforce in 1998, the business R&D workforce grew at a rate of more than 20 percent annually and by 2006 was about a quarter larger.[23] These impressive growth rates must be understood in the context of a low base period. As a share of the total working population, R&D workers grew from about 0.3 percent to about 0.7 percent.

The high-skill IT workforce grew by almost 70 percent between 1996 and 2000, but it has declined a little in the years since then, to about 52,000 in 2006. These occupations represented 1.1 percent of the labor force in 1996 and 1.5 percent a decade later, after a peak in 2000 at 1.7 percent.[24] Males dominate both the R&D and high-skill IT workforces, accounting for about 90 percent of the former and 80 percent of the latter.

We have relatively little data about the flows into and out of the R&D workforce. R&D occupations are more likely to be filled by those who have graduate degrees than are other occupations. The research-intensive universities of Hong Kong awarded 1,351 research-based graduate degrees in S&T fields in 2007, more than twice the 640 awarded in 1997. Census data for 2001 and 2006 suggest that roughly a third of R&D workers in Hong Kong during those years received their highest degrees outside of the SAR, but only about 10 percent reported that they had lived outside of Hong Kong 5 years earlier. The vast majority of these R&D positions are filled by Hong Kong-born permanent residents.[25]

High-skill IT workers have less need for formal education than R&D workers. Only about half of IT/computer professionals and associate professionals, as classified by the 2006 census, held graduate degrees, compared with approximately three-quarters of R&D workers.[26] The late 1990s' boom seems to have prompted a large number of Hong Kong residents – more than 60,000 – to acquire some advanced training, although not necessarily a degree, in the IT/computing field. This figure is far greater than the growth in the high-skill IT workforce, which was less than 20,000. Undergraduate degrees in IT programs funded by the UGC numbered about 1,400 in 1997, hit a peak of almost 2,000 in 2003, and fell back to about 1,750 in 2006.[27] The cumulative 10-year total is just over 17,000 undergraduate degrees, but the timing of this production does not match the growth of the high-skill IT workforce, which peaked in 2000.

As it does for the R&D workforce, in-migration plays a modest role for the high-skill IT workforce. Only a fraction (well under 5 percent) of this group in either of the two census years were not permanent residents of Hong Kong. Of those who held at least an undergraduate degree, about a third received their highest degree abroad.

We have limited data on the demand for R&D and high-skill IT workers. Annual surveys suggest that the market for R&D talent is fairly tight.

In 2004, about 20–25 percent of Hong Kong establishments identified the lack of qualified personnel as an important barrier to innovation, although this number declined to 10–15 percent in 2006.[28] R&D jobs also pay quite well, with 40 percent of workers in the field reporting incomes of over HK$40,000 in 2001. However, this share shrank to just under 30 percent in 2006. IT workers also suffered income declines in the 2001–2006 time frame, as one might expect given the decline in employment.

Looking forward, if the growth rate from 1998 to 2006 were to be maintained until 2012, Hong Kong's R&D workforce would double. Assuming that the overall working population of Hong Kong continues to grow at the same rate as in the past (1 percent annually), R&D's share would grow to almost 1.3 percent. This figure approximately matches the 2005 R&D share of the working populations of Beijing and Singapore, and is slightly less than Taiwan's 1.5 percent. We have no internationally comparable figures for the high-skill IT workforce.

4. Expanding the pool: Policy options

Our review of the data reveals that Hong Kong has been accumulating human resources, both general and specialized to S&T, at a relatively rapid rate in the past decade. The demand for this talent has kept pace reasonably well overall, although income data suggest some softness in recent years. There is evidence of a talent shortage only in a few specific categories, notably (in our review) for managerial, executive and R&D positions, and (anecdotally) for certain engineering specialties (which our data are too coarse to confirm). These trends reflect the private decisions of individuals and employers, but also the attention that has been paid to human resource issues, such as higher education and immigration, by Hong Kong's public policymakers. "The government," wrote the Commission on Strategic Development in 2007, "strongly believes that a larger pool of talent will increase our competitiveness, make Hong Kong more prosperous, attract more capital, and create more jobs" (CSD, 2007, p. 5).[29]

Although the trends point in the right direction and the gap seems to be closing, Hong Kong's talent pool still trails the leading "global cities" with which it is often compared. The region appears to be keeping up, broadly speaking, with its competitors in East Asia, but these competitors are developing quite rapidly themselves. The dynamic nature of international competition and the aspirations of the Hong Kong population suggest that policymakers ought to continue to focus on the human resources agenda as a key part of Hong Kong's economic development strategy. Merely maintaining the momentum of the past decade will present challenges, and these challenges would be heightened if the government seeks to accelerate the human capital formation process.

We discuss in this section a variety of options that the government might employ to bolster higher education and high-skill immigration. We want to be quite clear that our endorsement of any specific proposals, much less a full-fledged "supply-push" approach, would depend on their being embedded in a coherent broader package that incorporates demand considerations as well. Particularly for specialized fields, an intensive supply push would be risky without an equally intensive commitment to complementary policies that would facilitate job creation in these fields. A systematic approach of this sort requires careful policy coordination.

4.1. Higher education

Traditional universities have a central role to play in any human resource development policy. They also present distinct challenges for policymakers. Their capital facilities, especially for S&T fields, are expensive, long-lived assets. Faculty hiring also represents a long-term commitment. Academic programs tend, therefore, to acquire an inertia that insulates them somewhat from shifts in labor market demand and student interests. Hong Kong's demographics sharpen the challenge; the number of high school graduates in 2020 is projected to be just half that of today (Olsen and Burges, 2007). There are, of course, non-traditional modes of delivering higher education, often to non-traditional students, that can fill many human resource needs. Hong Kong has relied heavily on these in recent years, as we discuss at the end of this subsection.

The eight UGC-funded institutions, especially the four that rank in the global top 200 list for 2008 compiled by *The Times*,[30] are the most important sources of elite talent for Hong Kong. Their prestige constitutes a capital asset that must be preserved. Dilution of quality is worth worrying about. But such concern can easily lead to excessive conservatism in a dynamic environment that calls for continual change.

We are not implying that Hong Kong's higher education system has fallen into this trap. Hong Kong University of Science and Technology's (HKUST) founding just 17 years ago, for instance, was a bold move that has paid off splendidly. It ranked 39th on *Times 2008* list. A very different kind of shift in the higher education system is currently in prospect.

The UGC-funded institutions will move in 2012 from a 3-year to a 4-year undergraduate degree, "a daunting task," in the words of UGC deputy secretary general Kesson Lee. Nearly 1,000 new academic staff are being hired in preparation for this shift.[31]

As we noted in the previous section, the number of graduates from UGC-funded institutions has not grown very much in the recent past. For undergraduate degrees, the growth rate has been only about 0.7 percent per year over the past decade. While expansion of the graduating class would create numerous additional challenges – physical, fiscal, and managerial – on top of those faced in moving to a 4-year system, we suggest that it be given careful consideration. An additional 4,000 graduates per year (on a current base of about 16,000) would require the growth rate roughly to triple for a 10-year period.

If we imagine such a program being carried out by 2020, UGC-funded institutions could, in principle, enroll almost half of Hong Kong's shrinking population of 18-year-olds. It may reasonably be argued that reaching this deep into the pool of high school leavers would compromise quality. Students from the Mainland are an obvious alternative. They currently make up about 8 percent of undergraduates at UGC-funded institutions, a figure which might need to triple or quadruple under this scenario. That would mean raising or eliminating the 20 percent quota on non-local enrollment, which was raised from 10 percent only in 2006.

Given the size of the Mainland population, this approach seems likely to address any concern about student quality. Hong Kong's universities would have to compete for the top Mainland students with the rest of the world's great universities. Hong Kong's universities have great strengths, such as prestige, proximity, linguistic commonality, and post-graduation career opportunities, to draw upon in such a competition. The SAR government may want to encourage stronger linkages between Hong Kong and Mainland universities in order to facilitate recruitment and improve the educational experience.

Some Mainland students may need public financial support in order to attend Hong Kong universities, and Hong Kong taxpayers may resist providing it. One justification for such subsidies would be the future contributions that many of these students will ultimately make to the Hong Kong economy. In fact, subsidies to Mainland students might be made conditional upon postgraduation work in Hong Kong. Hong Kong has begun to permit non-local graduates of its accredited universities to stay for up to a year after finishing their degrees to look for a job. If they remain employed, they can stay on indefinitely. Using the university system as a mechanism to facilitate skilled migration seems to be an effective policy for Australia and (less explicitly) the US. This strategy allows potential immigrants to become socialized, and it provides a screening tool for the receiving country as well.

Some 2 percent of the undergraduate population of UGC-funded institutions currently comes from outside Hong Kong and China. Although this group may be more difficult to recruit and to induce to work in Hong Kong after graduation, it has other attributes that may justify a more aggressive outreach effort. To the extent that Hong Kong's economic advantage lies in linking China to the rest of the world, and vice versa, these students may contribute by creating long-lasting social connections with their local and Mainland colleagues. These students also strengthen the cosmopolitan outlook of Hong Kong's great universities, which is essential to a high-quality educational experience in an age of globalization.

With respect to the disciplinary composition of degree recipients, our view is generally consistent with the current policy, which is to let it be driven largely by student demand. But, as we have noted, lags and path dependencies help to shape demand, especially in the S&T fields, which are more capital-intensive than others. (In other words, students may choose their majors on the basis of the available facilities, rather than their true interests or opportunities.) Hong Kong has been rapidly expanding its capital investment for research training, as evidenced by the more than doubling of research-based graduate degrees awarded in 2007 compared with 1997 and the announced addition of some 800 (40 percent) more places. These moves should go some considerable way toward alleviating shortages in R&D personnel. Some measures targeted at drawing women into S&T fields might also be valuable, given the extreme gender imbalance in the R&D workforce. Continued expansion of S&T fields at the graduate level would make sense if Hong Kong's broader economic strategy emphasizes R&D-intensive industries and the R&D service sector itself. Undergraduate enrollments have not changed nearly as much as graduate enrollments, but the growth of research capacity overall suggests that a future surge in undergraduate demand might be accommodated fairly easily if it should materialize.

Much of the growth in the provision of undergraduate education in the past decade has occurred without UGC funding, although the providers are often arms of UGC-funded institutions. Self-financed subdegrees, in particular, have experienced "phenomenal" growth (Education and Manpower Bureau, 2006). What seem to be in short supply are so-called "top-up" programs that allow individuals with some undergraduate education to complete their degrees. During the 2007–2008 academic year, for instance, there were nearly nine times as many places available in self-financed subdegree programs as in self-financed degree programs.[32] Government policy seems to envision the subdegree as a terminal degree, but that is not how many subdegree holders and subdegree seekers perceive it.

It seems sensible to encourage Hong Kong's colleges and universities, both public and private, to expand their capacity to meet this emerging, self-financed demand. The government (or appropriate educational organizations) may need to articulate the regulatory framework that governs such matters as accreditation, transfer of credits, and experiential learning.[33] Such a framework will be valuable to non-traditional students who return to school later in life, as well as to university-age students who first move through subdegree programs. In the absence of opportunities to pursue "top-up" degrees in Hong Kong, some ambitions will be stifled, while many of those with the means to do so will go abroad.

4.2. Immigration

Immigration is the second major source of human resources that Hong Kong might draw upon. Siu et al. (2005) note the city's historical dependence upon this source, originating as a "space of flow" with porous boundaries. They also stress that "global cities" like New York and London rely on "continuous circulation" of population to remain economically vibrant. Like these cities, Hong Kong has long been relatively open to long-distance migration from the rest of the world. Circulation between Hong Kong and its nearby hinterland on the Mainland, by contrast, was disrupted for many decades and has only recently begun to bear a faint resemblance to that of, say, London to the rest of England. The more aggressively Hong Kong seeks to build up its human resource base, the more rapidly the balance of flows from China on the one hand, and from the rest of the world on the other, is likely to tip toward China. This shift will test popular attitudes about immigration.

Before we consider immigration policy options with respect to the Mainland and the rest of the world, we will briefly mention a third potential source of human resource inflows, the Hong Kong diaspora. We estimated in the previous section that 6,500 Hong Kong-born undergraduate degree holders returned to the SAR annually between 2001 and 2006. Olsen and Burges (2007, p. 7) estimate that about 9,000 students from Hong Kong begin undergraduate work abroad each year, although this number may well grow as more seek "top-up" degrees abroad. These figures are a fraction of the total stock of skilled Hong Kong expatriates in OECD countries, which was estimated to be over 290,000 in 2000 (Docquier and Marfouk, 2004).

Undoubtedly, many of these people are long-settled and have no interest in returning to Hong Kong. But perhaps some do or could be enticed to. A number of other Asian economies, including Taiwan and South Korea, have benefited greatly from return migration, including migrants who had been away from their home country for decades (Saxenian, 2006). Return migration to these countries has been encouraged by public policy. To our knowledge, the SAR government has no policy toward the diaspora and very little information about it. Data gathering would be a minimal first step in order to allow the government to assess whether more extensive measures, such as outreach and the provision of incentives to return, might be worthwhile.[34]

Hong Kong's immigration policy toward the Mainland must be viewed in light of its troubled history. The ideological conflict between the British colonial government and the People's Republic, and the vast gulf in living standards between Hong Kong and the Mainland, has inevitably left a complex legacy. One aspect of this legacy is an Immigration Department whose historic core competence is exclusion. Public opinion, too, has historically been skeptical about immigration from the Mainland and occasionally alarmed about it. (Of course, many Hong Kong people have family ties to the Mainland that temper these views.)

The notion that human resource considerations ought to shape immigration policy toward the Mainland is a relatively new idea, which has taken some time to penetrate these obstacles. A series of programs in the late 1990s and early 2000s that focused on particular occupations and industrial sectors drew in (at most) only a few hundred people annually, just a few percentage points of Hong Kong's skilled immigration from the rest of the world, admitted under the GEP (Siu et al., 2005, p. 130 and Table 3.3.6). In 2003, the Mainland-oriented programs were consolidated into the ASMTP, which is not restricted by occupation or sector. ASMTP has grown to about a quarter the size of GEP now.[35]

ASMTP is a demand-driven approach. A Hong Kong employer who wishes to hire a skilled Mainlander must demonstrate that the new hire's skills, knowledge, or experience are not readily available in the Hong Kong labor market. The compensation package must be commensurate with local norms. The vast majority of applications under the ASMTP are approved, about half of them for academic positions. The new program for non-local graduates of Hong Kong universities has a similar design.

The design is a good one for general human resource development, facilitating access to the large Mainland talent pool without crowding Hong Kong residents out of opportunities. However, it is possible that demand to immigrate is somewhat suppressed, as a result of the history described above. Continued positive experience with skilled immigration from the Mainland ought to provide an impetus toward further growth, in the context of a public that is "less resistant, but not yet supportive," in the words of Immigration Department deputy director, David Chiu.

The Hong Kong government added a modest supply-push component to skilled immigration policy in 2006, the Quality Migrant Admission Scheme (QMAS). Like comparable programs in Canada, Australia, and elsewhere, individual applicants earn "points" toward admission on the basis of attributes such as age, education, work experience, and language. The initial response to this opportunity was far below the quota of 1,000 per year; only 322 applicants were accepted in the first year and a half. About 60 percent of those accepted were Mainlanders. The point scheme was revised in early 2008, in part to dispel the perception that the QMAS applied only to Nobel prizewinners, Olympic medalists, and entertainment superstars, such as the pianist, Lang Lang.

Another potential deterrent for applicants under the QMAS is the discretionary nature of the decision. In "point" systems abroad, surpassing a set threshold earns admission. In Hong Kong, the Immigration Department and any expert advisors whom it chooses to engage decide each case individually. In principle, this set-up may allow this supply-push policy to be coordinated with the broader economic development strategy, as we have urged in this chapter. In practice, the program has so far been too small to make a difference. If a broader policy aimed at improving innovation in Hong Kong is enacted, the QMAS could provide the foundation for an associated human resource thrust, particularly if the expert advisors are knowledgeable about the strategy and given more authority over admission decisions.

The GEP is a demand-driven immigration policy that applies to skilled immigrants from other countries. The structure of the program is similar to that of the ASMTP, except that the applicant is the employee, rather than the employer. Although admissions under this policy have generally grown over time, they declined in 2002 and 2003, suggesting that applications do indeed reflect demand. Some interviewees expressed concern that broad quality of life considerations, such as education for school-age children and environmental pollution, were a deterrent to some potential in-migrants.

Some interviewees also suggested that the Hong Kong government should make a greater effort to promote in-migration opportunities at the world's talent centers. One characterized the current policy as "sit and wait." We have not looked into this issue in any detail but would offer the suggestion that the major responsibility for any such promotional effort should not be vested in the Immigration Department. It performs its core functions with exceptional efficiency. Promotion of opportunities is, to some extent, at odds with these functions. Invest HK, which promotes inward investment in the SAR, is a logical alternative agency for this assignment.

This brings us to our final point in this section. An effective human resource development policy depends on adequate coordination of a variety of agencies and organizations that perform very different functions, but whose collective efforts have a profound effect on Hong Kong's talent pool. If Hong Kong departs further from its tradition of "positive non-intervention" by pursuing an innovation-oriented economic strategy, the policy coordination challenge will be heightened. Matching future human resources, produced by higher education institutions and immigration policies, to future demand, produced by investments in R&D and the like, may require the government's central administration to be strengthened.

5. Conclusion

Hong Kong has made impressive strides in building up its talent base over the past decade. Growth in both general and specialized human resources, as measured by undergraduate degree holders in all fields and in S&T fields, has been strong, and the expansion of the R&D working population has been exceptional. The SAR fares reasonably well in international comparisons. A sound basis has been laid for continued movement in these directions.

Demand for highly skilled people has, if anything, lagged somewhat behind supply, especially in the past few years. We would therefore caution against moving forward on an aggressive stand-alone "supply-push" policy. However, accelerating the human capital formation process may well be a crucial component of a broader economic development strategy.

Hong Kong would face both challenges and opportunities in pursuing this objective. Demographic and institutional forces will constrain the domestic supply. Hong Kong will need to find ways to reach out more assertively to the rest of the world for talent, especially to Mainland China. Its excellent higher education system ought to be a valuable resource in this effort, and it can build on its nascent achievements in immigration policy toward trained professionals as well.

Notes

1. Ehrlich and Murphy (2007, p. 2) provide a helpful definition of human capital: "an intangible asset, best thought of as a stock of embodied and disembodied knowledge, comprising education, information, health, entrepreneurship, and productive and innovative skills, that is formed through investments in schooling, job training, and health as well as through research and development projects and informal knowledge assets." This definition encompasses both formal and tacit knowledge embodied in human beings.
2. Unless otherwise noted, the data in Section 3.1 are drawn from the Hong Kong By-Census (including a special tabulation provided by the Census and Statistics Department on 10 November 2008) and refer to holders of a 3- or 4-year degree granted by a college or university. As Olsen and Burges (2007) note, Hong Kong data sometimes fail to distinguish between attendance in a degree program, receipt of a subdegree, and completion of an undergraduate degree.
3. The 1996 by-census did not report a figure for degree holders. Averaging the figures from the 2001 and 2006 by-censuses, we estimate that 90.7 percent of those in the working population who were reported as attending degree programs in 1996 ultimately received a degree. We use this figure in the text whenever degree holding in the 1996 population is discussed.
4. UGC degree figures were provided by Ms Jenny Yip of UGC on 10 November 2008.
5. This estimate is calculated from the UGC figures referenced above and from *Hong Kong as a Knowledge-Based Economy* (Census and Statistics Department, 2007, p. 51).
6. These figures were provided by the Immigration Department on 29 October 2008. The 2007 figure was well over 26,000, and arrivals in 2008 through September were on pace to surpass that figure by several thousand.
7. As we discuss below, Hong Kong recently instituted a program that will permit mainland-born graduates of Hong Kong universities access to the Hong Kong labor market. This channel was too small to be accounted for in the text, but may grow rapidly in the future.
8. Some of the local undergraduate degree recipients and returnees may go to graduate school, be unemployed, work without pay in the home, or be retired. However, those admitted under the GEP and ASMTP are expected to be working as a condition of their admission. The *Hong Kong Yearbook* estimates emigration, including all levels of education, to be about 10,000 per year. However, recent graduates who go abroad, presumably temporarily, may not be included in this figure.

9. Data on income, employment, and unemployment in this paragraph are drawn from the GHS. Similar data in later subsections also rely on this source.
10. The survey does not distinguish between those who attended degree programs and those who ultimately received a degree.
11. The Hong Kong Education and Manpower Bureau is expected to complete soon a labor force projection based on a major modeling effort.
12. These figures should be treated with caution. In particular, we would stress that the degree to which they capture economic regions of differing sizes and compositions as well as variations in definitions and methods. Some observers also suggest that data on degree holders in Mainland China are inflated.
13. We include in these totals the census categories of life sciences, physical sciences, mathematics and statistics, computing, health, architecture, environmental protection, construction, engineering (civil, structural, mechanical, marine, production, industrial, chemical, electrical, and electronic), and textiles and clothing technology.
14. The ratio of degree holders to degree attendees in the 2001 and 2006 censuses is slightly higher for S&T degrees (about 93 percent) than for all degrees (91 percent). We use this figure to estimate the 1996 population of degree holders, which was not included in that year's by-census.
15. UGC categories included in this total are medicine, sciences, and engineering.
16. Some portion of the difference is surely accounted for by differences in definitions between census and UGC data.
17. To elaborate briefly, the 5,500 persons referred to in this paragraph are in the resident population, rather than the working population. If we assign 80 percent of them to the working population (as in Figure 3.2), we have a 4,400 person inflow. The total annual growth in the population is 9,300 and the annual inflow from UGC-funded institutions is 9,000, leaving a gap between total growth and the two sources of inflow of about 4,000. However, the definition of fields in census and UGC data does not match precisely, so measurement problems may explain some or all of the gap.
18. This replacement process includes both males and females. Females accounted for about 40 percent of the HOS working population in 2006, and the female share has grown steadily since 1996 across all subcategories of this population.
19. GHS shows more balanced growth among the three components of HOS. Since 2001, the population of managers and administrators as measured by the GHS has grown by more than 100,000, whereas the census shows a growth of only about 12,000.
20. *Hong Kong Social and Economic Trends, 2001–2007*; *2007 Report of Salaries and Employee Benefits*.
21. High-skill information technology (IT) workers cover personnel working in areas of IT/software development; IT sales; telecommunications and networking; IT education and training; general IT management; field support; systems programming; database; and IT security.
22. Census and Statistics Department, *Research and Development Statistics of Hong Kong*, various years. Data before 1998 are not publicly available. From 2000 to 2006, the growth rate was over 15 percent per year.

23. Between 2000 and 2006, the rate of growth in business R&D personnel was about 30 percent per year.
24. Census and Statistics Department, *Hong Kong as a Knowledge-Based Economy* (2007).
25. It is very important to note that the definition that we provided to the Census for these R&D workforce data covers "Physical, Mathematical and Engineering Science Professionals" only. We therefore use only proportions and urge caution in interpreting these data.
26. Please note that the census data used here may not match precisely the definition of high-skill IT worker used in the reports from which we derive other figures in this section. We use the occupations listed in the text: "IT/computer professionals and associate professionals."
27. Census and Statistics Department, *Hong Kong as an Information Society*, 2001 and 2007 editions.
28. *Annual Survey of Innovation Activities in the Business Sector*, 2001–2006.
29. In his November 2007 visit to Hong Kong, Chinese Premier Wen Jiabao also called upon Hong Kong to boost its efforts in this area.
30. The universities listed are Hong Kong University (#26), HKUST (#39), Chinese University (#42), and City University (#147). "University Top 200 in Full," *Times Online*, 9 October 2008.
31. David Mowery's chapter in this volume addresses the question of how these additional staff might be allocated across fields.
32. *Hong Kong Yearbook, 2007*. Combining self-funded programs with publicly funded programs, there are about twice as many places for subdegrees as for degrees.
33. Vivek Wadhwa's chapter in this volume addresses continuing education in the workplace.
34. Douglas Fuller's chapter in this volume argues that the chip industry in Hong Kong would benefit from an effort to reach out to the Hong Kong diaspora in Silicon Valley.
35. An additional 200–300 mainland passport holders who have been living abroad for at least a year are admitted annually under the "Relaxed Scheme," which adheres to the same conditions as ASMTP. And, about the same number are admitted under a similar policy that applies to mainland graduates of Hong Kong universities who left Hong Kong after graduation (CSD, 26 January 2007, p. 8).

References

Abel, Jaison R., and Todd M. Gabe, "Human Capital and Economic Activity in Urban America," Federal Reserve Bank of New York, Staff Report no. 332, July 2008.

Acemoglu, Daron, "Technical Change, Inequality, and the Labor Market," *Journal of Economic Literature* 40:7–72 (2002).

Berger, Suzanne, and Richard K. Lester, *Made by Hong Kong* (New York: Oxford University Press, 1997).

Commission on Strategic Development (CSD), "Policies and Schemes for Attracting Mainland and Overseas Talents," 26 January 2007.

Docquier, Frederic, and Abdesalam Marfouk, "Measuring the International Mobility of Skilled Workers (1990–2000) – Release 1.0," World Bank working paper 3381, 2004.

Education and Manpower Bureau, "Review of the Post-Secondary Education Sector," March 2006.

Ehrlich, Isaac, and Kevin Murphy, "Why Does Human Capital Need a Journal?," *Journal of Human Capital* 1:1–7 (2001).

Enright, Michael J., Edith E. Scott, and Ka-mon Chang, *Regional Powerhouse: The Greater Pearl River Delta and the Rise of China* (John E. Wiley, 2005).

"Guidelines for the Medium- and Long-Term National Science and Technology Development Program, 2006–2020," reprinted in Levin Institute, *Industrial Innovation in China* (2006).

Hu, Mei-Chih, and John A. Mathews, "National Innovative Capacity in East Asia," *Research Policy* 34:1322–1349 (2005).

Lucas, Robert E., Jr., "On the Mechanics of Economic Development," *Journal of Monetary Economics* 22:3–42 (1988).

Olsen, Alan, and Peter Burges, "Ten Years On: Satisfying Hong Kong's Demand for Higher Education," Strategy for Policy and Research in Education, 30 June 2007.

Romer, Paul M., "Endogenous Technological Change," *Journal of Political Economy* 98:S71–102 (1990).

Saxenian, Annalee, *The New Argonauts* (Cambridge: Harvard University Press, 2006).

Siu, Helen, Richard Wong, and David Faure, "Rethinking Hong Kong's Human Resources and Competitiveness," Hong Kong Institute for Humanities and Social Science, October 2005.

Tsang, Donald, "A New Direction for Hong Kong," 2007–2008 Policy Address, 10 October 2007.

US Department of Agriculture (USDA), Economic Research Service, "International Macroeconomics Dataset," available at <http://www.ers.usda.gov/Data/Macroeconomics/>, accessed 11 November 2008.

U.S. National Science Board, *Science and Education Indicators 2008* (Washington: U.S.N.S.B., 2008).

4
Workforce Development in Hong Kong

Vivek Wadhwa
Duke University and Harvard University

In many industrialized nations, key stakeholders frequently discuss whether the country has adequate talent to successfully compete in an era of rapid globalization. In the US, political and business leaders frequently express alarm at the aging engineering and scientific workforce. Declining enrollments in scientific fields in US universities are juxtaposed against rapidly accelerating graduation rates of engineers and scientists in China and India. This disparity often leads to predictions of severe shortages of science and engineering workers in the US and a loss of US competitive edge unless science and engineering graduation rates can be lifted. Globalization is also causing disruptions in the US workforce, with the outsourcing of high-skilled jobs to India and China. Protectionists say the solution is to raise trade barriers and prevent outsourcing to save quasi-technical, skilled white-collar jobs.

Ironically, stakeholders in Hong Kong are having a similar debate. They feel Hong Kong needs to expand its graduation rates of engineers and scientists to provide the brainpower required for a push into R&D-intensive industries. There are also calls for Hong Kong to increase its investment in industrial and academic research, with the assumption that more investment equates to innovation. This research-driven innovation would supposedly enhance competitiveness. However, Hart and Tian's chapter illustrated that as Hong Kong universities steadily increased output of science and technology graduates, demand for the majority of scientific talent actually softened. The only talent shortages they found were at the managerial and executive levels and for R&D positions in select engineering specialties.

The Global Engineering and Entrepreneurship project at Duke University has been researching the effect of globalization on the engineering profession and on US competitiveness. We have explored such topics as engineering education in India, China, and the US; the globalization of innovation and R&D; the impact of immigrants on the US economy; and how globalization is impacting intellectual property creation and entrepreneurial activity in the US.

We found that regularly cited graduation statistics for India and China were misleading and based on faulty comparisons. Our interviews with the executives of technology and engineering companies engaged in outsourcing R&D to India and China revealed that their primary motivation in moving operations abroad was not a shortage of engineers but lower cost and the proximity to growth markets. Furthermore, we found that the quality of engineering education in China and India was substandard for the purposes of many Western businesses.

Despite educational deficiencies, the outsourcing of R&D to India and China is gaining substantial momentum. In Hyderabad, India, companies like Satyam Computer Services and Hindustan Computers Limited are designing the interiors of luxury jets, in-flight entertainment systems, collision control/navigation control systems, fuel inverting controls, and other key components of jetliners for American and European corporations. In New Delhi, Indian scientists are discovering drugs for GlaxoSmithKline. In Pune, Indians are helping design bodies, dashboards, and power trains for Detroit automakers and soon may develop entirely outsourced passenger cars. In Bangalore, Cisco Systems, IBM, and other US tech giants have made the Indian city their global base for developing next-generation telecom solutions used in tomorrow's intelligent cities.

China is already the world's biggest exporter of computers, telecom equipment, and other high-tech electronics. Multinationals and government-backed companies are pouring hundreds of billions of dollars into next-generation plants to turn China into an export power in semiconductors, passenger cars, and specialty chemicals. It is lavishly subsidizing state-of-the-art labs in biochemistry, nanotech materials, computing, and aerospace. Within 15 years, China expects to be producing commercial aircraft to rival those of Boeing.

China is investing massively and has a top-down focus on achieving these feats. The Indian government is largely playing politics with its education system and has invested relatively little in infrastructure, education, and basic research. Yet India appears to be moving ahead of China to become a global hub for advanced R&D in several industries despite the lack of government investment in workforce development or infrastructure. In trying to understand why India has been able to pull ahead, we learned that the Indian private sector has found a way to overcome deficiencies in its education system through innovative programs of workforce training and development. These have transformed workers with a weak educational foundation into R&D specialists.[1]

With an aging population and low fertility rates, Hong Kong requires more than new children to maintain its global competitiveness. Unlike Mainland China, Hong Kong does not have significant problems with its engineering or science education programs, or with the supply of talent. The solutions do not lie in the schools and universities. The solution is to upgrade the skill sets of the existing workforce. Therefore, we believe that the Indian experiences in upgrading its workforce may be relevant for Hong Kong. We also believe, in combination with this, that Hong Kong can build on its tradition of business and trading – which is often called entrepreneurship. They key is to teach small businesses how to become mid-sized and large enterprises. In the following sections, we will detail what we learned from our interviews with Hong Kong companies in different sectors and provide observations about the skills available.

1. Workforce development in Hong Kong and engineering talent

To gain an understanding of the skill needs and workforce development practices in Hong Kong, and the landscape for engineering talent, we interviewed the following domestic, multinational, and Chinese companies in a diverse assortment of industries and spoke to several experts. Here we present the views of the interviewees. Note that most of the comments we have highlighted were echoed by a wide cross-section of people interviewed.

1.1. Opinions of some experts

Experts we interviewed include Alice (Miu Hing) Au, Managing Partner with Heidrick & Struggles; Thomas Goh, Partner with The Gallup Organization; Professor Alfred Ho, Executive Director of Academy of Management Consultancy; Professor Otto C. Lin, China CEO of Nansha Technology Enterprises Ltd; Charlie Y. Shi, Managing Partner of Omaha Capital China; Dr Thomas S. K. Tang, Executive Director of Global Institute Tomorrow, and Cheah Chin Teik, President of Chin Teik Consulting Ltd.

We summarize their comments in the passages below. We believe these provide meaningful insights.

- While the Hong Kong Science Park may have some advantages in terms of intellectual property, Hong Kong does not have the ecosystem for R&D. R&D in China is centered in Beijing and to some degree Chengdu. Hong Kong managers tend to have greater breadth and scope and general managerial perspective because they have had greater exposure. However, Hong Kong has fewer senior managers in non-financial sector areas than Singapore. There are very few Mainland Chinese general managers at senior levels, and Mainland Chinese managers tend to be more functionally oriented and lack non-technical skills such as marketing and sales. The pipeline of Chinese managerial talent is improving but will take 10–20 years to develop.
- During the past 5 years, Hong Kong has reinvented itself as a gateway to China in financial services, hospitality, and other service industries. Young Hong Kongese are more globalized, well-connected, and speak better English than their parents, but are also less willing to venture out than their parents were. Hong Kong companies traditionally do not spend on training and development because of historically high turnover. Hong Kong companies have become aware of the need to invest in training during the past 10 years. Shanghai companies are catching up to Hong Kong in terms of management knowledge, but they still lag behind in their ability to apply management knowledge. Hong Kong managers have had greater exposure to international business practices and global competition, and are more adaptable, confident, and cosmopolitan than their Chinese counterparts. In addition, Hong Kong managers have experienced both periods of growth and setbacks and are therefore able to respond to a greater range of business scenarios. A typical Shanghai manager would focus on execution versus collaboration, while Hong Kong managers have stronger project management and communication skills. Hong Kong has a 10-year lead in terms of managerial capabilities, but the People's Republic of China (PRC) will catch up quickly.
- Managerial training and development grew as part of business development in Hong Kong. Training and development units generally report to Human Resources (HR), and it is uncommon to have a director of training and development, reflecting the general level of importance assigned to training and development by senior management. SMEs (small and medium enterprises) play a particularly large role in Hong Kong, which has 680,000 SMEs registered for a 3.5 million workforce. Companies are easy to start, and most SMEs are

family-owned businesses, which invest little in training and development. Foreign-owned, large companies, and large government and non-governmental organizations, are most likely to invest in training. However, high turnover makes many companies reluctant to invest in training. Few companies are engaged in R&D in Hong Kong, and most companies adopt existing technology and focus on low-cost production and reliability. Many local engineering graduates do not enter technical production or engineering as Hong Kong has shifted to a service economy, and the services sector is able to absorb this talent. Most engineer graduates prefer to work in product sales and marketing than in manufacturing.

- There is a large supply of technical talent on the Mainland that is less expensive than Hong Kong graduates. China's entry into the World Trade Organization (WTO) has forced many business sectors to open up, and Mainland Chinese managers are learning quickly and are better informed than they used to be. China has many bright people, many who have pursued overseas studies. Hong Kong young people have more international connections and language advantages but are more complacent than their Mainland counterparts.
- There are eight universities in Hong Kong and five with strong engineering programs, which produce 10,000 graduates each year. The number of foreign students from Mainland China is increasing but most do not stay because of opportunities and family ties in the Mainland as well as the language barriers. Many engineering graduates enter design, sales and services, and product development functions. Hong Kong has strengths in product development, logistics, and supply chain management, and is the base for many company headquarters. Professional, legal, and financial services sectors are booming. In order for R&D to happen in Hong Kong, it requires a core group of corporate industrial R&D people. The majority of Hong Kong's business people are traders and middlemen involved in commerce and services, and they do not see the value of investing in innovation. Hong Kong business people are very entrepreneurial but are short-term oriented.
- There is no meaningful R&D in Hong Kong, and most R&D in China is limited to copying technologies and business models. Family-owned companies in Hong Kong are highly dependent on their founders and children in their quality of management. There is no culture of training, and in-depth processes do not exist. Top-tier scientists do not return to China and prefer the working environment

of the US. Investing in R&D in China is risky because there is a dearth of the managerial skills required for successful R&D.

- Hong Kong's education system is focused on examinations and obtaining qualifications versus applying information, and does not encourage independent thinking. Training is similarly viewed as a stepping-stone for advancement, and many employees will leave once they are trained. SMEs in Hong Kong are driven by profits and therefore invest little in training and development. As second and third generations of families that own SMEs obtain education in the West, they are bringing new managerial ideas to Hong Kong companies. Hong Kong graduates tend to have greater sophistication and ability to apply knowledge than their Mainland Chinese counterparts, but Hong Kong graduates also often lack communication and language skills.
- Hong Kong has strengths in financial services, sales and marketing, and hotel and tourism. Because the economy is doing well, Hong Kong does not feel the need for innovation. Many multinational corporartions (MNCs) in Hong Kong invest significantly in training, and many individuals pursue part-time studies. Chinese companies are generally not interested in training and development, and are focused on building and selling companies. However, individual demand for training is high as individuals know that they need to obtain knowledge and capabilities in order to compete.

1.2. Some data points from leading companies based in Hong Kong

1.2.1. Agilent Technologies

The best workforce development practices we observed in Hong Kong were at US-based Agilent Technologies. Agilent produces test and measuring equipment for electronics, life sciences, and chemical analysis. The company's Asia headquarters and sales and marketing are based in Hong Kong, where it has been operating for 20 years. Agilent has not pursued R&D in Hong Kong because the company has been unable to secure tax incentives or government subsidies. Agilent executives said that hiring technical people in Hong Kong is a challenge. Many top students are choosing to go to business school rather than engineering school, and many top engineering graduates opt for non-engineering jobs, particularly in the finance sector.

Agilent has sizeable R&D operations in Beijing and Shanghai, however. The company recruits from the top Mainland universities and says that Chinese graduates are great individual contributors and technically brilliant. However, these Chinese graduates lack project management, product marketing and product definition skills. Chinese-born students tend to be stronger academically and work harder, while Hong Kong students are more open to ideas, vibrant, and creative. The company sponsors a 90-day on-boarding program for most new hires. This includes company orientations as well as training online, in the classroom, and on-the-job training for technical and soft skills. The program also includes a 2-week overseas visit of factories. All new employees are assigned a mentor in the HR system, and these relationships last for 1 year. In the electronics group, engineers receive 4–5 weeks of ongoing development each year including 1 week of formal technical training, 10 days of technical online training, and 1–2 weeks of soft skills training. Managers receive 3–4 weeks of ongoing training each year. Agilent has a program for first-time managers, which includes classroom and online training modules received over the course of 6–9 months.

1.2.2. Bank of China (Hong Kong)

Bank of China is a Mainland-based bank competing with domestic and foreign banks in Hong Kong. To remain competitive in this environment, Bank of China has focused on employee development to a degree that is unusual in China. The bank employs 200,000 globally and 13,000 in Hong Kong, including 156 managers in Hong Kong. Approximately 10 percent of these managers come from the Mainland, and the remainder are from Hong Kong. Of the bank's 17 departments, only 4 are headed by managers sent from the Mainland headquarters to Hong Kong. The bank's officer training program annually recruits 200–300 fresh graduates from the top universities in China. In 2007, the program recruited 39 students from Hong Kong. Salaries are significantly higher in Hong Kong than in Mainland China.

Training and development efforts have focused on increasing productivity which improved from HK$890,000 per employee in 1988 to HK$1.3 million in 2004. The bank spends 1.8 percent of total personnel expenses on training programs. Training expenditures have remained flat for the past 3 years and are determined by the company board of directors. General staff are required to undergo 30 hours of mandatory training each year in areas such as internal control, risk management, anti-corruption, and money laundering. The bank built a dedicated training center in Hong Kong in 2003. The training center has 30 full-time staff and offers 1,100 courses in technical, soft, language, computer, and vocational skills.

In Hong Kong, the bank conducts 80,000 man-days of training each year. That equates to roughly a week of training per employee. Actual training per staff employee averaged 28 hours in 2007. Managers are required to undergo 40 hours of mandatory training each year. In 2007, managers averaged 68 hours of training. Participants in the Management Training (MT) program undergo 2 months of full-time classroom training led by internal trainers and managers. The MT program is for 3 years, and begins with an 18-month rotational program. MT program trainees are also assigned mentors.

1.2.3. China Netcom Hong Kong

China Netcom Communications (CNC) is a state-owned telecommunications provider. CNC is a full-service provider in Mainland China but focuses on wholesale customers (international operators) in Hong Kong. CNC has never hired fresh graduates and always hires experienced employees (usually in their 40s) in order to reduce training costs. For junior level staff, the company looks for candidates with 3 years of experience. The company has found that despite positive economic conditions, it often receives very senior applicants for relatively junior positions. CNC offers no training in Hong Kong.

Because of the large number of unemployed college graduates, attracting talent is not a major issue for state-owned companies. Employees who join state-owned companies still expect to remain for life, and state-owned enterprises (SOEs) do not face competition to retain employees. Employees who have worked for SOEs enjoy security and lack the international experience required to obtain jobs with international companies.

1.2.4. Clover Group International Limited

Clover Group is an intimate apparel manufacturer founded in Hong Kong in 1956 by Andy Lau, a Chinese immigrant. The company has 10,000 employees in China and 13,000 worldwide. Clover's 250 employees in Hong Kong are involved in merchandising, product development, human resources, and finance. The company says labor costs in Mainland China are one-tenth of those in Hong Kong. Operational managers come from Mainland China while general managers are mostly from Hong Kong. Turnover in sales and management is relatively low, while attrition in human resources is higher due to a booming market for HR professionals. Turnover is also high in Clover's merchandising group, measuring 25–30 percent annually. To improve retention, Clover sponsors work–life balance and recreational activities such as classes for stress management. It also sponsors social responsibility initiatives and has invested in enhancing managerial skills.

The company rents a building devoted to training, and workers receive 40 hours of technical training to develop multiple skill sets. This requirement is part of the ISO 9000 certification process. The company's existing managers often have good technical skills but lack supervisory and people skills. The company recently signed a US$4 million 3-year training and development contract with an external provider, targeted at managers, and last year completed US$1.4 million of training. The first phase focused on Hong Kong and included 20+ supervisors. The second phase will be rolled out in China and the third phase in Cambodia. Clover Group has conducted competency assessments and focus groups to identify skill gaps and training needs, has undertaken succession planning, and has enhanced its on-boarding program and social responsibility initiatives. The company would like to require at least 20 hours of training for managers.

1.2.5. Citrix Systems

Citrix Systems, a multinational software company based in Fort Lauderdale, Florida, builds software for hardware virtualization and secure remote access software products. Citrix employs 5,500 people in 100 offices around the world including 50 sales and support staff in Hong Kong. Company executives say that Hong Kong suffers a shortage of skilled human resources personnel. Most junior and middle-level human resources managers have moved up through administrative functions and have traditionally reported to finance. While there have been efforts to set up human resources Master's degree programmes and to offer strategic human resources management courses, Citrix management believes that human resources management capabilities of the Hong Kong workforce are still limited. Human resources issues are not a priority on the agenda of managers who are busy and overworked. The learning environment in Hong Kong companies is highly transactional, and employees are given tasks with little coaching or mentoring. In recent years, executive coaches have become very popular as some companies, particularly in the finance sector, have recognized the need to train and develop managers.

1.2.6. Dragonchip Ltd

Dragonchip was founded in 2001 with four employees and currently has 13 employees. The company designs semiconductor applications for use in remote controls and low-power hand-held devices. The company does design in Hong Kong and leverages outsourcing to manufacture in China. Dragonchip only hires experienced employees. Training and development is primarily done through on-the-job training. The founders spend approximately 20 percent of their time training employees.

1.2.7. Esquel Group

Esquel Group is the world's largest premium shirt maker and is a vertically integrated company. Esquel's businesses include cotton seed R&D, cotton cultivation, spinning, weaving, assembly, accessories, and a small but expanding retail brand. The Group has production facilities in China, Malaysia, Vietnam, Mauritius, and Sri Lanka, and a network of branches servicing key markets worldwide. Esquel manufactures for some of the world's best known brands, including Tommy Hilfiger, Hugo Boss, Brooks Brothers, Abercrombie & Fitch, Nike, Lands' End, and Muji, and major retailers such as Marks & Spencer, Nordstrom, and Jusco. The company has 47,000 employees, with 33,000 based in China and Hong Kong. Esquel Group says it is the leader in its apparel industry in terms of training and development. It spends less than US$2 million annually, or 1 percent of payroll costs, on training. The company plans to increase this figure in the near future.

First-time supervisors receive 5 days of training, and supervisors receive 2 weeks of ongoing development each year. Supervisors manage approximately 70 employees. They are trained in communication skills, technical skills, grievance management, meeting organization, and other managerial skills. First-time middle managers undergo 3 days of intensive training on different aspects of management, including communication, presentation, and reporting. Many, but not all, senior managers undergo training and development. The general manager of the accessory group, for example, spends 10 days each year on training and development. The company is aiming to build Esquel University by 2011 and to make it compulsory for all senior managers to spend time training. The company does not have a systematic executive education program. However, 4 years ago the company began to provide education sponsorships for executive MBA programs and technical training such as Six Sigma.

1.2.8. Gold Peak Industries (Holdings)

Gold Peak is a family-owned, publicly listed company in Hong Kong. The company's core products are batteries, GP lighting, crystal products, and high-end loudspeakers. The company is not developing any significant new technologies. They are involved in product design but not R&D. Gold Peak recruits engineers from China, back-office finance, human resources, and IT staff from Hong Kong, and product designers from Hong Kong and the United Kingdom. The company has 20 design employees in Hong Kong and 15 design employees in London. The company recruits radio frequency engineers from universities in the Guangdong area. Among these universities is Southeast University located in Guangzhou and close to Gold Peak's office in Shenzhen. Local graduates are more willing to come to Shenzhen that graduates from Beijing, Shanghai, or other regions.

The company has faced challenges in recruiting high-quality candidates in both China and Hong Kong. As manufacturing companies have moved from Hong Kong to Southern China, fewer Hong Kong graduates have entered manufacturing, and Hong Kong lacks middle managers in their 30s and 40s with technical know-how and experience in manufacturing. Senior managers in their 50s and 60s have moved up the corporate ladder at Gold Peak through a series of manufacturing positions. However, Gold Peak managers feel that today most Hong Kong graduates seek to enter financial services or other sectors.

The company says it would hire more Hong Kong managers if they were available because of their high levels of integrity, flexibility, business sense, and language skills. Gold Peak's experience is that Mainland Chinese managers perform well as supervisors and junior managers but have trouble functioning effectively as senior managers and top executives. The company launched a management trainee program in 2005. The program recruits trainees from top universities in Hong Kong, including Hong Kong University and the University of Science and Technology, and from top universities in Southern China. The program is aimed primarily at developing local talent in China to take advantage of lower salaries, and to fulfill business needs on the Mainland. The program also aims to counterbalance the lack of candidates in Hong Kong. During 2008 Gold Peak hired five graduates in Hong Kong and 25 in China as part of this program. The company's staffing levels in Hong Kong are stable, but turnover in China is much higher. The company pays at slightly above the median and does not have any specific retention initiatives. Gold Peak executives say the company spends 1 percent of payroll expenses on training and development and has no plans to increase this investment.

1.2.9. iASPEC Services

iASPEC is an IT consulting and software services firm headquartered in Hong Kong. The company was founded in 1998, launched its Shenzhen operations in 1992, and currently has 300 employees. The company recently moved to Zhuhai from Shenzhen due to rising costs. iASPEC says it has stopped hiring project managers from Hong Kong and hires Chinese project managers who are less expensive and more effective because they are able to spend more time with employees, have a better understanding of client needs, and have superior language and cultural skills. iASPEC's team leaders and resource managers are based in China, and its technology architects are based in Hong Kong. Company executives say that, while China produces large numbers of engineering graduates, they are not comparable in terms of skills or quality to Hong Kong graduates. Challenges with Mainland China graduates include reluctance to disagree with authority, fear of confrontation, inability to compromise, and lack of willingness to work out differences.

1.2.10. Noble Group

Noble Group is a commodity supply chain management and trading company headquartered in Hong Kong. The company was founded in 1987 by a UK businessman and is one of the few global commodity businesses in Hong Kong. The company has grown from 1,000 employees in 2004 to over 10,000 employees in 2008. These numbers include 5,000 crew members on ships, 3,000 employees in facilities, and 2,000 employees in offices. The pool of experienced talent in commodities trading is small, and Noble Group competes with investment banks and hedge funds for top talent. The company uses strategies such as looking for less experienced talent and providing greater pay incentives to attract talent. Recruits for its training programs are undergraduates and Masters graduates with 0–2 years of experience. Only 20 percent of trainees are from Hong Kong, China, and Singapore.

The company says that Chinese graduates have greater ambition, drive, and willingness to work than their Hong Kong counterparts. Both Hong Kong and Chinese graduates have weaker English language and cultural skills than their US, European, and Latin American counterparts. The company has also developed a customized executive education program in conjunction with INSEAD. The program is targeted at all managers, from first-line manager and above, in the age range of 28–54. First-line managers are typically in their late 20s. The program consists of three 1-week modules held in the two INSEAD campuses in France and Singapore, and in a Hong Kong location. The program culminates with trainees presenting an actual action proposal to the executive board. In addition, the company conducts management skills training sessions in which 12 selected managers from across Europe spend a day of training, discussing case studies and participating in role play exercises using real situations and led by an external consultant. A similar program is held in Argentina, and the company plans to implement a similar program in Asia.

1.2.11. ON Semiconductor

ON Semiconductor is a Motorola spin-off headquartered in Phoenix, Arizona. ON has 10,000 employees, 5,700 of whom are located in Asia, including 200 in Japan, 300 in China, 1,500 in Malaysia, and 89 in Hong Kong. ON does not hire fresh graduates in Hong Kong, and hiring is concentrated at the strategic and senior management level. The company says that Hong Kong still has good engineering schools, but the popularity of engineering and the quality of applicants has decreased significantly in recent years due to the lack of innovative and challenging engineering positions in Hong Kong. Top graduates enter financial services rather than engineering fields. The company also believes that Hong Kong managers tend to be more cosmopolitan, are less internally focused, and have greater exposure to MNC business practices and Western education than their Chinese counterparts.

1.2.12. Prosten Technology Holdings

Prosten was founded in Hong Kong in 1988 as a hardware trading company. Prosten has revamped its business model several times to focus on artificial intelligence research and developing mobile music search systems for customers such as China Unicom. The company currently has HK$83 million in revenues and 170 employees. Prosten hires almost exclusively in China. The company had 70 managers in Hong Kong in 2000 and 2001 but currently has only 11. Prosten hires all of its managers from the Tsinghua University MBA program. Finance, legal, sales, and administrative staff are plentiful and easy to hire. However, competent technical managers are difficult to find in China. Chinese engineers are strong technically but lack the experience to manage complex projects with multiple teams and layers, according to Prosten executives.

Retaining technical staff is challenging. The company faces a significant threat in its mobile music search business from Baidu, Google, and Yahoo. These companies pay high salaries for top talent and are constantly hiring away engineers from smaller competitors. Hong Kong engineers have somewhat better engineering skills than their Mainland China counterparts. However, Hong Kong does not have the skilled labor to support complex software development. The region's emphasis on financial services means that, while Hong Kong has many data processing department support managers, it does not produce skilled software developers.

1.3. Engineering graduates – Declining quantity and quality?

Our interviews revealed that companies had difficulty hiring skilled engineering talent in Hong Kong for R&D tasks. This is consistent with the findings by Hart and Tian, which show that the expansion of supply of science and engineering (S&E) talent in universities has not met industry needs and has not resulted in a boom in R&D jobs in Hong Kong. Several business executives said that the technical quality of graduates from Hong Kong universities had actually declined over the last decade as admission standards had been lowered in an effort to increase the number of engineers. At the same time, executives we spoke with claimed that Hong Kong locals had lost interest in studying engineering or, if they had studied the subject, had preferred careers in finance upon graduation. While engineering seems to be less attractive to Hong Kong graduates, Mainland Chinese students are filling the void and the companies we interviewed had no problem locating R&D and more technical jobs on the Mainland while keeping HR, strategy, or business-oriented jobs in Hong Kong.

1.4. Where do the engineering graduates go?

Clearly Hong Kong has expanded its output of engineering graduates but they have not gone into R&D jobs. Globally, financial services has traditionally offered the highest salaries and attracted the best talent from various fields, including engineering. This has also appeared to be the case in Hong Kong. As a result, semiconductor companies such as ON Semiconductor do not hire fresh graduates to do circuit design in Hong Kong. Instead, these firms are expanding their R&D in India and China. Similarly, in the software industry, companies like Prosten say that R&D is not possible in Hong Kong due to the lack of software engineers. This forced Protsen to move R&D to the Mainland. At the same time, local companies say they have a very hard time attracting and retaining Mainland Chinese engineers to Hong Kong. These transplants see greater opportunities on the Mainland and prefer to return home. Hong Kong graduates could go to the Mainland for technical jobs, but considering the comparatively lower salaries it is not a surprise that many choose not to make the move. Most multinationals that hire engineers in Hong Kong hire them for jobs in sales and marketing rather than in R&D. In the United States, Japan, and Europe, engineers usually move into sales and marketing jobs at a later stage in their career development. Because this happens earlier in Hong Kong, it creates further disincentives for students to study engineering, which they consider they will be unlikely to use in any real capacity.

1.5. The issue: Demand not supply

Seeking to boost the output of engineering graduates does little to address the core problem of lack of demand for science and engineering graduates in Hong Kong. While the government had hoped that increasing the supply of engineers would spur job creation, in fact on the Mainland and in India the reverse has happened: a shortage of talent and a surplus of jobs (demand) fueled expansion in engineering graduates. The comparative supply advantage of the Mainland influenced Hong Kong companies to move more of the few existing science and engineering jobs there where better engineers could be hired more cheaply. The lack of good engineering jobs in Hong Kong also makes it harder to retain Mainland talent, who see less of an engineering career path in Hong Kong. As a result, Hong Kong engineering graduates who remain in technical disciplines often leave Hong Kong for the US, Europe, and Mainland China. When they decide to become entrepreneurs and start companies, they do so in other countries.

1.6. The missing middle – Hong Kong's declining competitive edge

As manufacturing companies have moved from Hong Kong to Southern China, fewer Hong Kong graduates have entered manufacturing. Senior managers in their 50s and 60s have moved up through the manufacturing ranks, while most Hong Kong graduates now enter financial services or other sectors. As a result, Hong Kong lacks middle managers in their 30s and 40s with technical know-how and experience in manufacturing.

Hong Kong's workforce has had significant advantages over China in managerial capabilities, including linguistic, cultural, and social skills, as well as in flexibility and creativity, and a stronger sense of rule of law. A greater exposure to Western business practices has also been very helpful as it has given Hong Kong managers a better footing for dealing with foreign partners. The resulting situation is a downward spiral of technology talent fed by the exodus of manufacturing, which, in turn, reinforces the loss of technology talent. Because there are fewer opportunities for engineering graduates, Hong Kong is unable to grow technical managerial talent. It is experiencing a hollowing out of its technical managerial talent, and Hong Kong's competitive advantage in managerial capabilities for technical or manufacturing enterprise is fast disappearing or, as in the case of most technology products, has already largely disappeared.

2. Recommendations for Hong Kong

To compete globally, Hong Kong needs to compete on its strengths – which include its global outlook, Western orientation, rule of law, intellectual property rights, entrepreneurial workforce, and professional management skills. The country needs to focus on strengthening those things that have provided it with advantages to date. Improvements are needed in the area of workforce development. Further, it is necessary to take workforce development beyond an orientation in financial services to new fields like advanced R&D. To turn Hong Kong into an R&D hub and compete more effectively with Mainland China for science and technology jobs and enterprise, Hong Kong must provide the incentives for students and workers in the existing workforce to acquire these skills. Augmentation of these skills, in turn, will help create more jobs in these areas. In other words, create the demand and facilitate education, rather than starting by trying to increase supply. The best way of doing this is to foster greater technology entrepreneurship and create hubs of technical activity.

Additionally, education and training in growing small business into medium-sized and large businesses will facilitate growth. And the country can import entrepreneurs from countries like the US, from where skilled workers are increasingly leaving due to frustration with US immigration policies.[2]

Notes

1. Wadhwa et al., 2007b. *Intellectual Property, the Immigration Backlog, and a Reverse Brain-Drain: America's New Immigrant Entrepreneurs*, Part III.
2. Wadhwa et al., 2008a. *How the Disciple Became the Guru.*

References

Gereffi, Gary, Vivek Wadhwa, Ben Rissing, and Ryan Ong. 2008. Getting the Numbers Right: International Engineering Education in the United States, China, and India. *Journal of Engineering Education* 97, no. 1: 13–25.

Wadhwa, Vivek, Gary Gereffi, Ben Rissing, and Ryan Ong. 2007a. Where the Engineers Are. *Issues in Science and Technology* (Spring).

Wadhwa, Vivek, Guillermina Jasso, Ben Rissing, Gary Gereffi, and Richard B. Freeman. 2007b. *Intellectual Property, the Immigration Backlog, and a Reverse*

Brain-Drain: America's New Immigrant Entrepreneurs, Part III. Marion Kauffman Foundation Working Paper. http://ssrn.com/paper=1008366.

Wadhwa, Vivek, Una Kim De Vitton, and Gary Gereffi. 2008a. *How the Disciple Became the Guru*. Marion Kauffman Foundation Working Paper. http://ssrn.com/paper=1170049.

Wadhwa, Vivek, AnnaLee Saxenian, Ben Rissing, and Gary Gereffi. 2008b. Skilled Immigration and Economic Growth. *Applied Research in Economic Development* 5, no. 1: 6–14.

Wadhwa, Vivek, AnnaLee Saxenian, Richard B. Freeman, and Gary Gereffi. 2009. *America's Loss Is the World's Gain: America's New Immigrant Entrepreneurs*, Part 4. Marion Kauffman Foundation Working Paper. http://papers.ssrn.com/sol3/papers.cfm?abstract_id=1348616.

5
On Reform of Hong Kong's Public Research Funding System

Chintay Shih
Tsinghua University

Shin-Horng Chen
Chung-Hua Institution for Economic Research

1. Introduction

The Hong Kong government used to adopt a laissez-faire policy to promote economic development in Hong Kong, but since 1997 promotion of R&D and innovation has been given high priority on the policy agenda (ITC, 2004; Liu, 2008). In order to come to terms with the rising economic strength of Mainland China, some in Hong Kong, from both the private and public sectors, feel that Hong Kong needs to develop its own innovation and technology capabilities in a few fields in order to become an "innovation hub" in the region (Baark and Sharif, 2006; Hills et al., 2004; Sharif and Baark, 2008). In so doing, Hong Kong may be able to capitalise on the manufacturing muscle in the Mainland , the Pearl River Delta (PRD) in particular.

Against this backdrop, the Applied Science and Technology Research Institute (ASTRI) was established in 2000 to conduct industry-oriented applied R&D using the Industrial Technology Research Institute (ITRI) in Taiwan and the Korean Advanced Institute of Science and Technology (KAIST) in Korea as its reference models. The Innovation and Technology Commission (ITC) was also established as the successor of CTI (Commission on Technology and Innovation) in 2000 to coordinate related policies to promote R&D and technological innovation in different sectors. Since then, R&D schemes and projects funded by the ITC have grown in number (see Section 3). However, questions remain as to the extent to which Hong Kong's more active innovation policies during the last decade have generated significant impacts (Liu, 2008).

This chapter is about Hong Kong's public research funding system, addressing the following main issue: given the goal to become an "innovation hub" in the region, how may the Hong Kong government reshuffle its public research funding system and innovation governance? In general, innovation governance has become an increasingly important issue and a key challenge for OECD (Organization for Economic Cooperation and Development) member countries. Addressing this issue requires developing the necessary institutional set-ups, procedures and practices for agenda setting and prioritisation, implementation and policy learning (OECD, 2005). Of course, innovation governance involves a lot of aspects and issues, but as far as this chapter is concerned, we are particularly interested in the way in which the Hong Kong government, the ITC in particular, administrates the R&D funding schemes and the way in which the research institutes interact with the funding agencies.

2. A snapshot of R&D in Hong Kong

This section highlights some stylised features of R&D in Hong Kong, though some more detailed statistics can be seen in the Appendix.

Hong Kong, though one of the original Asian Newly Industrialising Economies, has not been very active in R&D and technological innovation. Its R&D expenditures in 2006 amounted to HK$11.95 billion (about US$1.54 billion), accounting for 0.81 per cent of GDP. This percentage is relatively low by international standards, lower than its major neighbouring economies, Taiwan (2.58 per cent) and China (1.42 per cent) (see also TechMatrix Research Centre, 2008). However, a positive trend has been the growing momentum of the private (business) sector in R&D investment. The business sector accounted for 53 per cent of the total R&D expenditure in 2006, and has overtaken the higher education sector to become the major R&D performing sector since 2005. On the other hand, the government sector, including the public technological supporting institutions, has played quite a minor role as an R&D performer, with its R&D share being as low as about 2.08 per cent in 2006, though the government remains an important source of funds for R&D. Of note is the fact that despite a publicised commitment by the government to the stepping-up of R&D activities, the R&D expenditure invested by the government either fluctuates over time or, at most, grows at a modest rate (HKCSD, 2008a).

A close look at R&D expenditure in the business sector by industry sector suggests a predominant role played by the service industry. Hong Kong-based enterprises with R&D activities are mainly clustered in two broadly defined industry sectors, namely, (1) the wholesale, retail, and import and export trades, restaurants and hotels sector; and (2) the financing, insurance, real estate and business services sector. These two sectors contributed 58 and 24 per cent of the total R&D expenditure of the business sector respectively, followed at a distance by the manufacturing industry (6 per cent). However, an extra part of R&D for manufacturing may be hidden in the wholesale, retail, and import and export trades, restaurants and hotels sector because R&D activities in this sector are predominantly performed by trading firms with subcontract processing arrangements. As for the financing, insurance, real estate and business services sector, R&D activities undertaken by the constituent firms were mainly related to information technology (HKCSD, 2008b).

In addition, there seems to be a mismatch between public R&D and private R&D, especially in terms of the strategic areas selected by the ITC for the R&D Centre Programme, including automotive parts and accessory systems, information and communications technologies, logistics and supply chain management enabling technologies, nanotechnology and advanced materials, textiles and clothing, and Chinese medicine. The R&D expenditure in the business sector concentrates predominately on information technology and electrical and electronic engineering technology, with these two areas contributing 42.5 and 33.6 per cent of the total private R&D respectively. On the other hand, for such areas as Chinese medicine and nanotechnology, the R&D investment of the private sector is negligible.

The business sector in Hong Kong is engaged substantially in R&D outsourcing. In 2007, a total of HK$3,223.7 million was spent by this sector in R&D outsourcing, representing more than half the total business R&D expenditure. The wholesale, retail, and import and export trades, restaurants and hotels sector, as a whole, not only was involved substantially in R&D outsourcing but also spent 93.9 per cent of the total expenditure for outsourced R&D activities (HK$2,442.5 million) to parties outside Hong Kong. This type of R&D outsourcing also accounted for about 74.8 per cent of the total expenditure for outsourced R&D activities by the financing, insurance, real estate and business services sector, the second largest R&D performing business sector. In terms of the geographical and organisational patterns of the performing parties of the outsourced R&D concerned, intra-corporate cross-border network, especially within the PRD Economic Zone, is the dominant type of R&D outsourcing adopted by Hong Kong-based firms/establishments. This is consistent with the above-mentioned significance of Hong Kong-based trading firms with subcontract processing arrangements in the wholesale, retail, and import and export trades, restaurants and hotels sector.

3. Hong Kong public R&D funding

Since 1998, a few funding schemes have been set up under the auspices of the ITC to support different innovation activities, ranging from R&D (the Innovation and Technology Fund (ITF)), technology ventures (the Applied Research Fund (ARF)),[1] and design (the DesignSmart Initiative), to patent application (the Patent Application Grant (PAG)). Table 5.1 outlines some of the major funding schemes administrated by the ITC; among these schemes, this chapter is particularly concerned with the ITF, as well as the innovation governance relationship between the ITC and its umbrella R&D institutes. Figure 5.1 portrays the structure of the ITC's funding schemes.

The Innovation and Technology Fund (ITF), launched in 1999 with an injection of HK$5 billion, aims to support projects that contribute to innovation and technology upgrading in industry, as well as those essential to the upgrading and development of industry. The ITF can be considered as the flagship R&D initiative funded by the Hong Kong government, not only because of the sheer size of its budget allocation but also because of the wider coverage of its funding structure. The ITF has four programmes: the Innovation and Technology Support Programme (ITSP), the University-Industry Collaboration Programme (UICP), the General Support Programme (GSP), and the Small Entrepreneur Research Assistance Programme (SERAP). Of particular relevance to this chapter is the ITSP because the lion's share of the research institutes' funding comes from this programme. In addition, according to the ITC (2008), the funding approved under the ITSP amounted to 83 per cent of the whole ITF from its initiation to May 2008 (see also Table 5.2).

In 2005 the ITC adopted a new three-tier structure for funding proposals under the ITSP. Tier 1 involves the establishment of R&D centres to undertake projects in their respective technology areas, including: automotive parts and accessory systems; logistics and supply chain management enabling technologies; textiles and clothing; nanotechnology, and advanced material and information and communications technologies. Tier 2 involves the funding of project proposals submitted

Table 5.1 Description of the major funding schemes under the auspices of the ITC

Scheme	Description	Notes

The Innovation and Technology Fund (ITF)	• Launched in November 1999 with an injection of HK$5 billion.
	• To support projects that contribute to innovation and technology upgrading in industry, as well as those essential to the upgrading and development of industry.
	• Four programmes under the ITF
	  • Innovation and Technology Support Programme
	  • University-Industry Collaboration Programme
	  • General Support Programme
	  • Small Entrepreneur Research Assistance Programme.
		• As at the end of January 2009, a total of 3101 applications received requesting HK$15.7 billion funding; among them, 1285 (HK$3.8 billion) approved.
		• Most of the funded projects related to information technology (30%); electrical and electronics (24%); and manufacturing technology (15%).

The Applied Research Fund (ARF)	• A government-owned venture capital fund to support local technology ventures with commercial potential, with a capital of HK$750 million. • Administered by the Applied Research Council (ARC), a private company wholly owned by the government.	• As at the end of January 2009, 24 investments with funding of HK$392 million made. • The investment period of the ARF expired at end March 2005 and the Fund has ceased making new investments.
The DesignSmart Initiative	• Launched in June 2004, with HK$250 million. • To strengthen government support for design and innovation, and to promote wider use of design and innovation in industries to help them move up the value chain. • Two main elements: financing a design support programme and setting up the InnoCentre as a one-stop shop for a design cluster.	• As at the end of January 2009, a total of 302 applications received; among them, 202 (HK$106.5 million) approved.

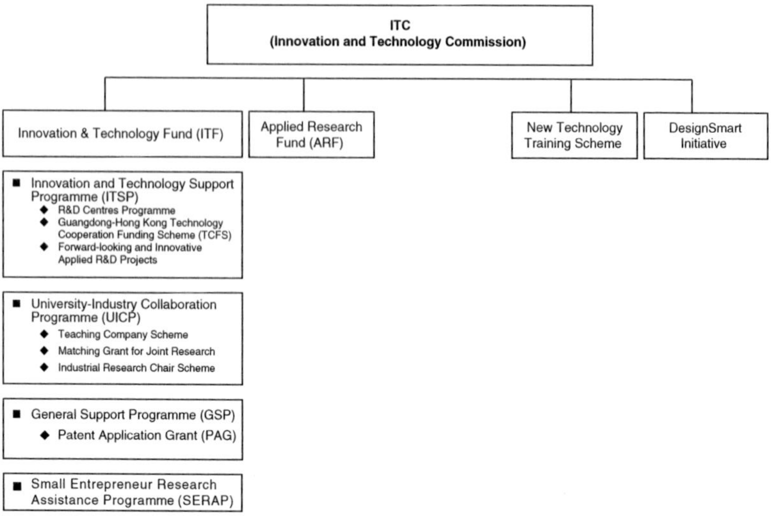

Figure 5.1 Structure of the funding schemes administrated by the ITC

under the Guangdong-Hong Kong Technology Cooperation Funding Scheme (TCFS). Tier 3 involves the funding of more forward-looking and innovative applied R&D projects (www.itc.gov.hk).

The R&D Centres Programme resulted from "New Strategy" released by the ITC in 2005, which consisted of two key initiatives. The first one was to identify the strategic technological areas to be actively promoted by the government. The underlying criteria for the selection of the focus areas included: (1) existing research capabilities of universities and other research institutes, (2) Hong Kong companies' competitive advantages, (3) industrial needs, and (4) market potential.

The second key initiative was to set up R&D Centres in selected areas to conduct applied R&D and to facilitate technology transfer from universities and research institutes to the business sector. The underlying aim was to support the further development of innovation and technology, with an emphasis on the five key elements of focus, market relevance, industry participation, leverage on the Mainland, and better coordination among different elements of the innovation and technology programme. In total, five R&D Centres were established in 2006 to drive and coordinate R&D efforts in the designated technology areas.

The six[2] R&D Centres are administrated by different host organisations, including the Hong Kong Productivity Council (automotive parts and accessory systems); ASTRI (information and communications

Table 5.2 Innovation and technology fund: Distribution of approved projects among different industrial sectors (as at 30/9/2008)

Industrial Sector	**Programme**								**Total**	
	Innovation and Technology Support Programme		**General Support Programme**		**University-Industry Collaboration Programme**		**Small Entrepreneur Research Assistance Programme**			
	No.	**$mn**	**No.**	**$mn**	**No.**	**$mn**	**No.**	**$mn**	**No.**	**$mn**
Biotechnology	77	236.3	5	2.8	37	62.9	23	22.6	142	324.4
Electrical and Electronics	225	1120.3	3	2.8	37	31.6	74	76.1	339	1,230.9
Environmental	19	44.2	1	1	8	19.6	12	12.5	40	77.3
Information Technology	139	709	8	9.6	42	49.1	139	144.2	328	912.1
Foundation Industries	135	667.4	4	3.2	46	27.9	13	12.6	198	711.0
Textiles/Clothing/Footwear	55	223.5	1	0.4	5	3.3	4	3.2	65	230.4
General (Cross Sectors)	3	37.7	84	92.9	–	–	1	0.9	88	131.5
Others	6	20.9	10	17.0	–	–	10	8.3	26	46.1
Total	659	3059.3	116	129.7	175	194.3	276	280.4	1226	3663.7

Notes: There may be a slight discrepancy between the sum of individual items and the total as shown in the tables owing to rounding.
Source: http://www.itf.gov.hk/eng/statistics/StatTable104View.asp?StatTypeId=104&StatId=517&StatCaption=Distribution+of+Approved+Projects+among+Different+Industrial+Sectors.

technologies); the University of Hong Kong, the Chinese University of Hong Kong and the Hong Kong University of Science and Technology (logistics and supply chain management enabling technologies); the Hong Kong University of Science and Technology (nanotechnology and advanced materials), the Hong Kong Polytechnic University (textiles and clothing), and the Hong Kong Jockey Club Institute of Chinese Medicine Limited (Chinese medicine). The designated functions for these centres include: (1) to conduct industry-oriented research; (2) to facilitate IP commercialization; (3) to provide technology and market intelligence; (4) to provide a platform for exchange of IT/technology; and (5) to promote technology development, transfer, and knowledge dissemination. However, according to our interviews in Hong Kong, at least some of these centres are more like project offices, with limited in-house R&D capacity and a limited number of staff members.

On top of that, ASTRI was established in 2000 to conduct industry-oriented applied R&D. ASTRI's research areas include photonics technologies, internet applications, wireless communications, and IC design. Its operating strategy is to transfer the technologies developed from its R&D projects to industry through licensing arrangements, contract research arrangements, and spinning-off new technology companies. ASTRI has a subsidiary company, the Hong Kong Jockey Club Institute of Chinese Medicine Limited (HKJCICM), which aims to promote and support the modernisation and further development of Chinese medicine in Hong Kong. The Hong Kong Jockey Club Charities Trust, which has a 50 per cent shareholding of the HKJCICM, has pledged to donate HK$500 million to fund the R&D activities managed by the HKJCICM, while ASTRI provides premises and supporting facilities to the HKJCICM and funds its recurrent operating costs.

Table 5.2 provides information on the breakdown of the approved projects under ITF by programme and industrial sector. The ITSP is the major programme funded by the ITF, accounting for 53.75 per cent of the approved projects and 83.5 per cent of the approved amount respectively. The broadly defined information and communications technologies (including information technology and electrical and electronics, shown in Table 5.2) are the field that is most funded across programmes under the ITF, followed by the so-called foundation industries and biotechnology.

Table 5.3 goes further to show the funding indicators of the major funding schemes, of particular interest to this chapter, under the auspices of the ITC. Over the period 2006–2008, among the schemes listed in Table 5.3, the ITSP has funded most projects. In terms of the R&D

Table 5.3 Funding indicators of the major funding schemes under the auspices of the ITC, 2006–2008

	2006 (Actual)	**2007 (Actual)**	**2008 (Estimate)**
ITSP$^{\Psi}$			
Applications received and processed	126	137	**137**
Projects funded and being monitored	117	116	**144**
TCFS			
Applications received and processed	62	51	**60**
Projects funded and being monitored	58	90	**103**
R&D Centres' projects$^{\varphi}$			
Automotive Parts and Accessory Systems R&D Centre			
New projects	—	9	**39**
Projects funded and being monitored	—	9	**48**
Hong Kong R&D Centre for Information and Communications Technologies			
New projects	16	35	**43**
Projects funded and being monitored	16	51	**93**
Hong Kong R&D Centre for Logistics and Supply Chain Management Enabling Technologies			
New projects	—	8	**13**
Projects funded and being monitored	—	8	**21**
Nano and Advanced Materials Institute			
New projects	—	5	**48**
Projects funded and being monitored	—	5	**53**
Hong Kong Research Institute of Textiles and Apparel			
New projects	—	13	**20**
Projects funded and being monitored	—	13	**33**

Table 5.3 (Continued)

	2006 (Actual)	2007 (Actual)	2008 (Estimate)
ASTRI projects*			
New projects	8	—	—
Projects funded and being monitored	32	25	11

Notes: $^{\Psi}$ The figures do not include applications submitted by or projects undertaken by the five R&D Centres and ASTRI, which are reported under the indicators "R&D Centres' projects" and "ASTRI projects".

$^{\varphi}$ All projects (including TCFS projects and feasibility studies) undertaken and/or monitored by R&D Centres are included. For the Hong Kong R&D Centre for Information and Communications Technologies (ICT R&D Centre), the actual figures for 2006 have been updated to include feasibility studies.

* The figures do not include projects undertaken by the ICT R&D Centre hosted by ASTRI, which are now reported under the new indicator "R&D Centres' projects". Separately, the actual figures for 2006 have been updated to include feasibility studies undertaken by ASTRI.

Source: Adapted from ITC, Controlling Officer's Report (2008).

Centres Programme, out of the five R&D Centres listed, the R&D Centre for Information and Communication Technologies, right from 2006, has made a quick start and has funded a number of projects, whereas the rest only began to gather momentum from 2007 with a limited number of projects funded. In the field of Chinese medicine, the HKJCICM, since its inception, has funded only 17 projects in collaboration with local universities and Mainland institutions (ITC, 2009).

The picture portrayed above may have something to do with the mismatch between public R&D and private R&D, especially in terms of the strategic areas selected by the ITC for the R&D Centre Programme. As discussed, the R&D expenditure in the Hong Kong business sector concentrates predominately on information technology (including information systems and technology, computer hardware technology, computer software technology, and communication technology) and electrical and electronic engineering technology, with these two areas contributing 42.5 and 33.6 per cent of the total private R&D respectively. This is equally true for the two major R&D investing sectors, namely the wholesale, retail, and import and export trades, restaurants and hotels sector, and the financing, insurance, real estate and business services sector. However, for areas such as Chinese medicine and nanotechnology, the R&D investment of the private sector is negligible. As a result, it is not surprising to see that the R&D Centre for Information and Communications Technologies has made rapid and significant progress in funding R&D projects.

4. Innovation governance of Hong Kong public R&D funding

4.1. Public R&D investment and portfolio

It is widely perceived that R&D investment and intensity in Hong Kong is quite low by international standards and when compared with its peer economies, the R&D intensity being as low as 0.81 per cent. In fact, despite a publicised commitment by the government to the stepping-up of R&D activities, the R&D expenditure invested by the government did not grow until 2005, and after that has had only a modest annual growth rate (given its low base) of less than 8 per cent. If Hong Kong's R&D investment keeps moving at the same pace, it will be unrealistic for Hong Kong to become an "innovation hub" in the region.

In contrast, China's R&D intensity reached 1.42 per cent of GDP in 2006 and the Chinese government intends to have R&D intensity reach 2 per cent by 2010. A closer look at the regional level within China reveals that for such localities as Beijing and Shanghai, R&D intensity in 2005 had reached as high as 5.5 and 2.3 per cent respectively; even Guangdong had an R&D intensity of 1.1 per cent (Blue Book of China's Regional Development, 2007, p. 12). Hong Kong's peer economies in Asia, Taiwan and Singapore, have both set a goal of raising R&D intensity to the 3 per cent level by around 2010.

More problematic is the way in which the Hong Kong government has used its R&D expenditure. In our views, the resource allocation of public R&D in Hong Kong is intrinsically short term. Although having strategic foci, the R&D initiatives funded by the Hong Kong government are by and large short-term oriented, dispersed, and reactive. These plans lack the long-term, integrated type of R&D that makes good "innovation policy" in a broad sense. For example, although ASTRI is positioned to conduct applied R&D, due to the low innovation capabilities of traditional electronics SMEs in the PRD, ASTRI has to develop technologies to an almost product-ready level, so that the recipient companies can apply the technologies developed directly to their production processes (Liu, 2008, p. 5). Also, even ASTRI, the flagship research institute, admittedly is mainly doing development work rather than genuinely forward-looking research because of the short time spans of the projects. In addition, not all of the R&D Centres supported by the ITSP are truly engaged in R&D activities, with some of them functioning simply as a project office. As a result, the R&D Centres Programme remains generally reactive, even though the R&D Centres currently have 5 years of funding.

In contrast, an important lesson from the OECD member countries has warned: "Budgetary practices often promote short-term thinking and in some cases undermine strategic, long-term policy making" (OECD, 2005, p. 8). As a result, efforts have been made in OECD countries to reduce fragmentation and create critical mass and excellence in the public research sector. Initiatives in this area include ensuring or strengthening block grant funding mechanisms to support longer-term research, especially in catching-up economies, or renewing support for infrastructure and research equipment in more advanced countries (OECD, 2008, p. 59).

Taking Finland as an example, the Finnish government has stepped up the model of centre of excellence by supporting a few Strategic Centres for Science, Technology and Innovation (otherwise known as SHOK). SHOKs provide a new way of coordinating dispersed research resources to meet targets that are important for Finnish business and society.

In the individual strategic centres, companies, universities and research institutes will work together to formulate a jointly agreed research plan. The plan will aim to meet the needs of member companies for practical application within a 5–10-year period. In addition to shareholders, public funding organisations will commit themselves to providing funding for the centres in the long term. Another good example is Singapore's "holistic" approach to the development of a leading centre of excellence in biotechnology, which does not just involve initiatives of funding (technology) policy but also requires a good combination of vertical programmes and horizontal programmes for the strategic areas (Vonortas, 2008).

To go a step further, if Hong Kong is to become an innovation hub in the region, the government needs to facilitate, in a few selected areas, the development of distinct capabilities and networking linkages that can prevail in the region or become a true centre of excellence. Indeed, with globalisation, support for clusters is also evolving with a view to creating world-class "nodes" to link to global innovation value chains rather than geographically bound clusters. Linkages and co-operation between regions both within and between countries are becoming more important (OECD, 2008). TechMatrix Research Centre (2008) has argued that with appropriate policy reform Hong Kong can leverage the "Extended Open Innovation" Business Model to become an innovation hub in the region. However, open innovation, first coined by Henry Chesbrough (2003) as an antithesis of closed innovation, cannot be reduced to just a better practice for the routine innovation process (Chen et al., 2008). Chesbrough (2003) has reminded us of the significance of architectures and systems and business models to the adoption of the open innovation model. He argues:

> Open innovation processes combine internal and external ideas into architectures and systems. Open innovation processes utilize business models to define the requirements for these architectures and systems… Open innovation explicitly incorporates the business model as the source of both value creation and value capture. This latter role of the business model enables the organization to sustain its position in the industry value chain over time… Open innovation treats spillovers as a consequence of the company's business model. These spillovers need not be a cost of doing business, they are an opportunity to expand a company's business model, or to spin off a technology outside the firm to locate a different business model. (Chesbrough, 2003)

We therefore would like to argue that if the ITC and the research institutes in Hong Kong continue to devote most of their resources and efforts to short-term industry-oriented R&D or problem-solving adaptive R&D, it will not be possible for Hong Kong to become an innovation hub in the region. It is imperative for public R&D in Hong Kong to have a more balanced R&D portfolio, at least by spending a certain portion of the public R&D investment on long-term strategic topics that may involve new architectures and systems and business models, in which some of the Hong Kong-based firms have a role to play. However, for this kind of R&D activity to take root in Hong Kong, in the research institutes in particular, it requires a more flexible governance relationship between the funding agencies and the research institutes, an issue to be discussed below.

There are concerns that the spillovers from R&D are not fully captured in Hong Kong, given the extensive offshoring and outsourcing of activities to the Mainland. There are in Hong Kong different views on this issue. One explicitly stated criterion for the ITC's major funding schemes, in particular the R&D Centre Programmme, is to leverage on the Mainland. As a result, many of the R&D initiatives orchestrated by the Hong Kong government end up by providing R&D results to the firms operating in the PRD. On the other hand, as a Shenzhen official put it: "How does Hong Kong benefit if R&D occurs in Hong Kong, but development and commercialization take place in Shenzhen".[3] This problem may become even more significant, if we take into account the fact that the fields targeted and/or identified by the government tend to be manufacturing-centric. The manufacturing sector accounts for a negligible part of Hong Kong's economy, even though part of the bourgeoning service sector and R&D in the major service sectors, as shown above, is manufacturing-related. Any hardware innovation resulting from the indigenous R&D investment in Hong Kong probably has to find its way to the Mainland for commercialisation, leading to the resultant job creation, if any, mainly in the Mainland. As a result, except for those elites engaged in R&D activities, most of the Hong Kong residents may not benefit much from the manufacturing-centric R&D investment, leading to a situation whereby the research institutes and the Hong Kong Science and Technology Parks become an "enclave" in Hong Kong.

In our view, the Hong Kong government should at least invest a certain portion of the public R&D expenditure in service innovation, particularly systemic service innovation, so that Hong Kong may go from strength to strength and serve as a "testbed" for brand new service innovation. In this way, Hong Kong may be better able to serve its residents as well as China by leveraging indigenous innovation and local needs, and eventually exporting services to China and the rest of the world. In fact, Hong Kong has managed to win the franchise bid to run an underground route in London. The fact that Hong Kong is such a metropolis may give rise to the type of sophisticated demand that spurs innovation. The success of the "Octopus Card", though based on Sony's technology, is a good example in this regard and its usage has expanded to cover Shenzhen. In addition, the plan to merge Hong Kong with Shenzhen to form a mega city in the future will be likely to pose challenging issues that will spur innovation. In fact, some professionals in Hong Kong endorse this idea of a "testbed" and suggest that the digital TV services may provide a good opportunity for Hong Kong.

It is worthwhile to note that systematic innovation of services entails large-scale transformation of the services as well as the goods involved. As demonstrated by den Hertog (2001), systemic service innovations require at least four elements in place: a new service concept, new client interface, new service delivery system and technological options, together redefining the role of the key actors involved and serving as a new value proposition. Therefore, when promoting systemic service innovations, the government needs to adopt a holistic and flexible approach, which will be different from that used in the promotion of manufacturing-centric R&D or technology-centric initiatives. In other words, different types of innovations can vary in nature and in key success factors. Therefore, public policy with regard to different types of innovations, particularly systematic service innovations, cannot be reduced merely to the provision of R&D funds, hence giving rise to the importance of innovation policy.

4.2. The funding mechanism and institutional arrangements

As vividly illustrated in the literature of the National Innovation System (NIS), the way in which the diverse innovation actors of a nation interact with one another within the NIS may be affected by the incentive schemes and institutional arrangements, and may thus lead to different innovation performances (Chang et al., 2004; Freeman, 1987; Lundvall, 1992; Nelson, 1993). Therefore, even though the R&D initiatives orchestrated by the ITC have expanded in number of funding mechanisms and institutions for pursuing R&D, the way in which the research institutes interact with the funding agencies remains an issue of particular concern. Below we would like to discuss some of the relevant issues raised in Hong Kong.

Though funded by different schemes, nearly all of the research institutes in Hong Kong are positioned to conduct industry-oriented applied research. The problem is that the funding schemes administrated by the ITC are by nature short-term oriented, dispersed and reactive, as discussed above. As a result, the research institutes are deficient in their R&D portfolio and short of the capital for making strategic investments. Taking ASTRI as an example, its research projects are supposed to meet three criteria at the same time: be innovative, have commercial value, and have a funding span from 1 to 1 1/2 years. As a result, ASTRI tends to be constrained to pursuing development work and/or "me-too" projects. This is compounded by the ITC's requirement of a basic industry contribution of 10 per cent for each project because the business sector in Hong Kong in general has a strong preference for short-term profitability.

In contrast, a research institute such as the ITRI in Taiwan, with financial support from the Department of Industrial Technology (DoIT) at the Ministry of Economic Affairs (MOEA), can propose and conduct long-term R&D projects, up to 4 or 5 years, though annual reviews for checking progress are still required. In addition, the DoIT's funding schemes for research institutes provide the latter with opportunities to conduct different types of research and/or strategic investment, ranging from pioneering technology research, to the building up of infrastructure required and large-scale R&D facilities in order to meet their long-term and strategic needs (see Table 5.4). More importantly, the assessment procedures can vary across different types of funding schemes. In particular, the assessment procedure for the Pioneering Technology Research Programme is conducted by the ITRI itself[4] in order

Table 5.4 The types of funding schemes for research institutes, administrated by the DoIT in Taiwan (Unofficial Translation)

1. Pioneering Technology Research Programme
2. Key Technology Base Program
(1) Forward-Looking R&D Program
(2) Key Technology/Product Program
(3) Infrastructure-Building Program
3. Large-scale R&D Facilities Program

Source: This study.

to provide appropriate room and flexibility for the formation of more creative projects.

For the R&D schemes under the ITF, the ITC explicitly requires industrial sponsorship for each project, not less than 10 per cent of the total project cost, with an aim to ensure industry orientation. Some local professionals, particularly those in the R&D Centres and universities, consider this requirement troublesome and too rigid. On the other hand, according to ASTRI, the ITC has shown some flexibility by allowing ASTRI to get an average of 10 per cent from a number of projects rather than a full 10 per cent for each project. In our views, the industrial sponsorship requirement may not be as unreasonable as it looks, but the problem is that the ITC's R&D funding schemes for research institutes are not as diversified as is the case with the DoIT in Taiwan. As a result, in response to such institutional arrangements, the research institutes in Hong Kong tend to focus on short-term-oriented development work and/or me-too projects. In order for the research institutes to have sound and balanced R&D portfolios, we suggest that the ITC should provide a wider variety of R&D schemes for the research institutes; some of the schemes still demand industrial sponsorship, while others allows the research institutes to involve strategic R&D, even without industrial participation, right from the beginning.

Some are uneasy with the regulation that approval from the Finance Committee of the Legislative Council is required for a project requesting more than HK$15 million from the ITF. A new development from October 2008 is the increase in the financial ceiling from HK$15 million to HK$21 million. However, even so, for such projects to kick off, the research proposals have to go through five panels, including internal review, industrial review, technology review, ITC review, and board of directors,[5] plus the Legislative Council. This process is really too lengthy and may cost the projects time-to-market lead time. It is recommended that, as with funding practices in many countries, the technology review committee should be given authority to make decisions, before reporting to the ITC for final approval.

Related to this, project management on the part of the ITC is often criticised. There is an impression that the ITC tends not to tolerate any changes in projects. Outputs from the R&D projects should be specified beforehand, especially in terms of what patents are to be filed. Any changes in the projects require heavy paperwork and approval from the ITC. It is essential that the ITC reduces administrative micromanagement. For example, changes in projects should be allowed up to a certain extent, and endorsed mainly by the review committee rather than always by the ITC, except for major changes. The way to evaluate the output, outcome and even impact of an R&D project is also an issue that needs to be dealt with. Some staff members of the R&D Centres are particularly concerned with what criteria are used for evaluating their performances. DoIT in Taiwan can even tolerate failures in some cases and evaluates the performance of the research institutes from a long-term perspective, especially regarding such a forward-looking and risky programme as the Pioneering Technology Research Programme.

The way in which the government manages the R&D Centre Programme also draws criticism. It seems to us that not all of the R&D Centres are truly engaged in R&D activities. To our understanding, except for the designated areas, the projects funded by the R&D Centres are not that different from those funded by some other major schemes administrated by the ITC. We suggest that upon completion of their project time span, the ITC should conduct an intensive evaluation of the performance of the individual R&D Centres. In particular, the ITC should review the R&D Centres' business plans, to be formulated by the centres and their major stakeholders, in order to determine whether these plans are in line with the ITC's long-term strategy. Based on this, the ITC may have to make a critical decision for the consolidation of the R&D Centres Programme in the near future. Consolidation may become even more necessary if one takes into account the fact of the mismatch between public R&D and private R&D, especially in terms of the strategic areas selected by the ITC for the R&D Centre Programme. Alternatively, a more ambitious policy for the ITC to adopt is to follow the Finnish model of SHOKs, or the holistic approach adopted by the Singapore government, in the development of regional centres of excellence in biotechnology.

Another issue concerns the shortage of dedicated professionals (technology managers) for the promotion of technology transfer on the part of the research institutions as well as the universities. According to ASTRI, its R&D personnel have to shoulder the work of R&D and the promotion of technology transfer at the same time. In Taiwan, not only do the research institutes have dedicated units for technology transfer and/or technology management, but the DoIT also provides the research institutes with the resources required through the Infrastructure-Building Programme. It is therefore advisable for the ITC to step up its efforts in this regard. In addition, Hong Kong can also take advantage of the training and support system built by the Association of University Technology Managers (AUTM) in the US and/or the

Association for University Research and Industry Links (AURIL) in the UK (Lee, 2006) to train and recruit the technology managers required.

5. Conclusions and policy recommendations

Hong Kong has begun to step up its efforts on R&D, but much remains to be done. Hong Kong has not been very active in R&D and technological innovation. However, a positive trend has been the growing momentum of the business sector in R&D investment. Of note is the fact that despite a publicised commitment by the government to the stepping up of R&D activities, the R&D expenditure invested by the government did not grow until 2005, and even after that the annual growth rate has been modest at less than 8 per cent.

A close look at R&D expenditure in the business sector by industry sector suggests a predominant role played by the service industry, namely, (1) the wholesale, retail, and import and export trades, restaurants and hotels sector; and (2) the financing, insurance, real estate, and business services sector. However, an extra part of R&D for manufacturing may be hidden in the wholesale, retail, and import and export trades, restaurants and hotels sector because R&D activities in this sector are predominantly performed by trading firms with subcontract processing arrangements.

In addition, there seems to be a mismatch between public R&D and private R&D, especially in terms of the strategic areas selected by the ITC for the R&D Centre Programme. The R&D expenditure in the business sector predominately concentrates on information technology and electrical and electronic engineering technology, with these two areas contributing 42.5 and 33.6 per cent of the total private R&D respectively. On the other hand, for such areas as Chinese medicine and nanotechnology, the R&D investment by the private sector is negligible. Public technology support organizations in Hong Kong seem to play a more active role in Hong Kong-based firms' R&D cooperation arrangements than is the case with R&D outsourcing. However, Hong Kong-based firms' cooperative partners, the higher education institutions (HEIs), seem to be more geographically dispersed, rather than mainly concentrated in Hong Kong.

Since the turn of the twenty-first century, R&D schemes and R&D projects funded by the ITC have grown in number, predominately through the ITSP under the ITF. The ITC has adopted a new three-tier structure for funding proposals under the ITSP since 2005. Tier 1 involves the establishment of R&D Centres to undertake projects in their respective technology areas. Tier 2 involves the funding of project proposals submitted under the Guangdong-Hong Kong Technology Cooperation Funding Scheme (TCFS). Tier 3 involves the funding of more forward-looking and innovative R&D projects.

The six R&D Centres are administrated by different host organisations but some of them are more like project offices, with limited in-house R&D capacity and a limited number of staff members. Of these R&D Centres, the R&D Centre for Information and Communications Technologies, right from its inception in 2006, has made a good start and has funded a number of projects, while the rest only began to gather momentum from 2007, and so far have funded only a limited number of projects. This may have something to do with the mismatch between public R&D and private R&D, especially in terms of the strategic areas selected by the ITC for the R&D Centre Programme.

The resource allocation of public R&D in Hong Kong is intrinsically short term. Although they have strategic foci, the R&D initiatives funded by the Hong Kong government are by and large short-term oriented, dispersed, and reactive.

The Hong Kong government should aim to invest a significant portion of the public R&D expenditure in service innovation, particularly the systemic service innovation, so that Hong Kong may go from strength to strength and serve as a "testbed" for brand new service innovation. However, different types of innovations can vary in nature and in key success factors. Therefore, public policy with regard to different types of innovations, particularly for systemic service innovations, cannot be reduced merely to the provision of R&D funds, hence giving rise to the importance of innovation policy.

Since the funding schemes administrated by the ITC are by nature short-term oriented, dispersed and reactive, the research institutes are deficient in their R&D portfolios and short of capital for making strategic investment. This is compounded by the ITC's requirement of a basic industry contribution of 10 per cent for each project because the business sector in Hong Kong in general has a strong preference for short-term profitability. In our views, the industrial sponsorship requirement may not be as unreasonable as it appears, but the problem is that the ITC's R&D funding schemes for research institutes are not as diversified as in the case of the DoIT in Taiwan.

Some are uneasy with the regulation that approval from the Finance Committee of the Legislative Council is required for a project requesting more than HK$15 million from the ITF, though the financial ceiling has now been increased to HK$21 million. The review process is also considered too lengthy and may cost the projects time-to-market lead time.

Related to this, project management on the part of the ITC is often criticised. There is an impression that the ITC tends not to tolerate any changes in projects. It is essential for the ITC to reduce administrative micromanagement.

The way to evaluate the output, outcome, and even impact of an R&D project is also an issue that needs to be dealt with. Hong Kong should learn from DoIT in Taiwan. DoIT can tolerate failures in some cases and evaluates the performance of the research institutes from a long-term perspective, especially regarding such a forward-looking and risky programme as the Pioneering Technology Research Programme.

The way in which the government manages the R&D Centre Programme also draws criticism. To our understanding, except for the designated areas, the projects funded by the R&D Centres are not that different from those funded by some other major schemes administrated by the ITC. We suggest that upon completion of their project time span, the ITC should conduct an intensive evaluation of the performance of the individual R&D Centres.

Another issue concerns the shortage of dedicated professionals (technology managers) for the promotion of technology transfer on the part of the research institutions as well as the universities. In Taiwan, not only do the research institutes have dedicated units for technology transfer and/or technology management, but the DoIT also provides the research institutes with the resources required through the Infrastructure-Building Programme. It is therefore advisable for the ITC to step up its efforts in this regard.

Above all, we would like to emphasise that if Hong Kong's R&D investment keeps moving at the same pace or does not manage to catch up with its neighbouring economies for years to come, it will be unrealistic for Hong Kong to become an "innovation hub" in the region. Also, if the funding mechanism and institutional arrangements continue to be based on an administrative mindset, Hong Kong's innovation governance will be undermined. To solve these problems from a long-term perspective, we suggest the Hong Kong government should follow the example set by its neighbouring economies and promulgate a Hong Kong version of "Science and Technology Basic Law" (Chang et al., 2004) which could facilitate the speeding up of public R&D investment with "additionality" and the sound development of innovation governance.[6]

Appendix: Major statistics of R&D in Hong Kong

Table A.1 R&D expenditure by performing sector, 2002–2006

Sector	R&D expenditure (HK$ million)				
	2002	2003	2004	2005	2006
Business	2505.8 ⟨33%⟩ [0.20%]	3545.1 ⟨41%⟩ [0.29%]	4590.3 ⟨48%⟩ [0.36%]	5621.6 ⟨51%⟩ [0.41%]	6287.4 ⟨53%⟩ [0.43%]*
Higher education	4800.7 ⟨64%⟩ [0.38%]	4796.2 ⟨56%⟩ [0.39%]	4707.3 ⟨50%⟩ [0.36%]	5085.0 ⟨47%⟩ [0.37%]	5410.9 ⟨45%⟩ [0.37%]*
Government	237.1 ⟨3%⟩ [0.02%]	207.5 ⟨2%⟩ [0.02%]	207.6 ⟨2%⟩ [0.02%]	215.2 ⟨2%⟩ [0.02%]	248.6 ⟨2%⟩ [0.02%]*
Total	7543.6 ⟨100%⟩ [0.59%]	8548.8 ⟨100%⟩ [0.69%]	9505.2 ⟨100%⟩ [0.74%]	10,921.8 ⟨100%⟩ [0.79%]	11,946.9 ⟨100%⟩ [0.81%]*

Notes: Figures in ⟨ ⟩ represent the percentages to total. The percentages in a year may not add up to 100 due to rounding.

Figures in square brackets represent the ratios to GDP. The GDP estimates are based on the data on expenditure-based GDP estimates at current prices released on 27 February 2008.

* Figures are subject to revision later on.

Source: Adapted from HKCSD (2008a).

Table A.2 R&D expenditure by source of funds, 2002–2006

Source of funds	R&D expenditure (HK$ million)				
	2002	2003	2004	2005	2006
Local parties					
Business sector	2655.3 ⟨35%⟩	3641.9 ⟨43%⟩	4538.3 ⟨48%⟩	5786.7 ⟨53%⟩	6304.1 ⟨53%⟩
Government sector	4736.6 ⟨63%⟩	4704.3 ⟨55%⟩	4467.8 ⟨47%⟩	4816.7 ⟨44%⟩	5151.4 ⟨43%⟩
Higher education sector	14.9 ⟨0.2%⟩	22.2 ⟨0.3%⟩	5.7 ⟨0.1%⟩	48.4 ⟨0.4%⟩	25.4 ⟨0.2%⟩
Others	7.4 ⟨0.1%⟩	1.6 ⟨§⟩	7.4 ⟨0.1%⟩	0.8 ⟨§⟩	2.2 ⟨§⟩
Parties outside Hong Kong	129.4 ⟨2%⟩	178.9 ⟨2%⟩	486.0 ⟨5%⟩	269.1 ⟨2%⟩	463.8 ⟨4%⟩
Total	7543.6 ⟨100%⟩	8548.8 ⟨100%⟩	9505.2 ⟨100%⟩	10,921.8 ⟨100%⟩	11,946.9 ⟨100%⟩

Notes: Figures in ⟨ ⟩ represent the percentages to total. The percentages in a year may not add up to 100 due to rounding.

§ Less than 0.05%.

Source: Adapted from HKCSD (2008a).

Table A.3 R&D expenditure in the business sector by industry sector, 2002–2006

Industry sector	R&D expenditure (HK$ million)				
	2002	2003	2004	2005	2006
Manufacturing	481.9 ⟨19%⟩	406.1 ⟨11%⟩	587.1 ⟨13%⟩	471.3 ⟨8%⟩	369.0 ⟨6%⟩
Wholesale, retail, and import and export trades, restaurants and hotels	936.3 ⟨37%⟩	1255.6 ⟨35%⟩	2310.8 ⟨50%⟩	2541.9 ⟨45%⟩	3676.1 ⟨58%⟩
Financing, insurance, real estate and business services	552.8 ⟨22%⟩	1493.7 ⟨42%⟩	1299.2 ⟨28%⟩	2184.4 ⟨39%⟩	1528.5 ⟨24%⟩
Others	534.8 ⟨21%⟩	389.7 ⟨11%⟩	393.1 ⟨9%⟩	424.0 ⟨8%⟩	713.8 ⟨11%⟩
Total	2505.8 ⟨100%⟩	3545.1 ⟨100%⟩	4590.3 ⟨100%⟩	5621.6 ⟨100%⟩	6287.4 ⟨100%⟩

Notes: Figures in ⟨ ⟩ represent the percentages to total. The percentages in a year may not add up to 100 due to rounding.
Source: Adapted from HKCSD (2008a).

Table A.4 Total expenditure for in-house R&D activities in 2007 by technology area by industry sector

	Technology area												
	Information technology					Electrical & electronics engineering technology*	Manufacturing technology	Biotechnology	Chinese medicine	Nanotechnology	Advanced materials technology	Others	Total(1)
	Information system and technology	Computer hardware technology	Computer software technology	Communication technology	Subtotal								
By industry sector													
Manufacturing	25.7 (4.3%)	47.6 (8.0%)	27.0 (4.6%)	8.1 (1.4%)	108.3 (18.3%)	201.0 (34.0%)	220.2 (37.2%)	4.3 (0.7%)	3.8 (0.7%)	3.9 (0.7%)	49.9 (8.4%)	0.1 (#)	591.5 (100.0%)
Wholesale, retail and import and export trades, restaurants and hotels	60.0 (2.2%)	126.1 (4.6%)	189.4 (6.9%)	273.2 (10.0%)	648.8 (23.7%)	1,410.8 (51.5%)	448.9 (16.4%)	31.6 (1.2%)	0.0 (0.0%)	45.1 (1.6%)	156.3 (5.7%)	0.0 (0.0%)	2,741.4 (100.0%)
Financing, insurance, real estate and business services	616.5 (28.4%)	189.6 (8.7%)	610.2 (28.1%)	146.9 (6.8%)	1,563.2 (72.0%)	399.3 (18.4%)	32.9 (1.5%)	65.5 (3.0%)	1.4 (0.1%)	14.8 (0.7%)	92.9 (4.3%)	2.2 (0.1%)	2,172.3 (100.0%)
Others	68.5 (12.5%)	25.3 (4.6%)	76.9 (14.0%)	80.9 (14.7%)	251.7 (45.8%)	26.1 (4.8%)	10.4 (1.9%)	238.5 (43.4%)	0.5 (0.1%)	0.1 (#)	0.9 (0.2%)	21.1 (3.8%)	549.4 (100.0%)
Total	770.7 (12.7%)	388.6 (6.4%)	903.5 (14.9%)	509.2 (8.4%)	2572.0 (42.5%)	2037.2 (33.6%)	712.5 (11.8%)	339.9 (5.6%)	5.8 (0.1%)	63.9 (1.1%)	299.9 (5.0%)	23.4 (0.4%)	6054.6 (100.0%)

Notes: (1) Figure includes expenditure for in-house R&D activities conducted by a local party for itself and/or for another organisation.
* Electrical and electronics engineering technology associated with (a) computer hardware (such as integrated circuits) was included in the area of computer hardware technology; (b) communications was included in the area of communication technology.
# Figure less than 0.05%.
Source: Adapted from HKCSD(2008b).

Table A.5 Total expenditure for outsourced R&D activities in 2007 by outsourced party by industry sector (HK$ million)

	Expenditure for outsourced R&D activities to local parties	Expenditure for outsourced R&D activities to parties outside Hong Kong	Total expenditure for outsourced R&D activities
By industry sector			
Manufacturing	22.6 (62.6%)	13.5 (37.4%)	36.1 (100.0%)
Wholesale, retail and import and export trades, restaurants and hotels	149.1 (6.1%)	2293.4 (93.9%)	2442.5 (100.0%)
Financing, insurance, real estate and business services	137.6 (25.2%)	408.8 (74.8%)	546.3 (100.0%)
Others	120 (60.4%)	78.7 (39.6%)	198.7 (100.0%)
Total	429.3 (13.3%)	2794.4 (86.7%)	3223.70 (100.0%)

Source: Adapted from HKCSD (2008b).

Table A.6 Total expenditure for outsourced R&D activities in 2007 by performing party and source of funds (HK$ million)

Type of organisation	Party performing R&D activities		Source of funds			
			Outsourced to local parties		Outsourced to parties outside Hong Kong	
Local parties						
Self-financed	–	(–)	310.1	⟨72.2%⟩	2607.5	⟨93.3%⟩
Government	11.9	⟨0.4%⟩	46.7	⟨10.9%⟩	8.7	⟨0.3%⟩
Public technology support organisations	28.1	⟨0.9%⟩	–	⟨–⟩	–	⟨–⟩
Higher education institutions	47.9	⟨1.5%⟩	0.7	⟨0.2%⟩	0.0	⟨0.0%⟩
Other business firms within an establishment's own enterprise group	136.4	⟨4.2%⟩	16.1	⟨3.8%⟩	8.9	⟨0.3%⟩
Business firms outside an establishment's own enterprise group	204.9	⟨6.4%⟩	54.3	⟨12.6%⟩	1.4	(#)
Others	0.1	(#)	0.0	⟨0.0%⟩	0.0	⟨0.0%⟩
Parties outside Hong Kong						
Other business firms within an establishment's own enterprise group	2025.9	⟨62.8%⟩	0.9	⟨0.2%⟩	101.2	⟨3.6%⟩
Business firms outside an establishment's own enterprise group	758.0	⟨23.5%⟩	0.5	⟨0.1%⟩	66.7	⟨2.4%⟩
Others	10.5	⟨0.3%⟩	0.0	⟨0.0%⟩	0.0	⟨0.0%⟩
Total	3223.7	⟨100.0%⟩	429.3	⟨100.0%⟩	2794.4	⟨100.0%⟩

Notes: – Not applicable.
Figure less than 0.05%.
Source: Adapted from HKCSD (2008b).

Table A.7 Distribution of establishments with R&D activities in 2007 by the types and location of their cooperation arrangements

Whether having cooperation arrangements on R&D activities with other organisations/Type of cooperation organisation[1]	**No. of establishments having undertaken R&D activities**	**Location of cooperation organisation**[1]					
			Mainland China and Macao				
		HK	**Pearl River Delta (PRD) Economic Zone**[2]	**Pan-PRD Region**[3] **other than PRD Economic Zone and HK**	**Other regions**	**Places outside HK, the Mainland of China and Macao**	**Overall**
Having cooperation arrangements on R&D activities with other organisations	**1339** [27.5%]						
Government		43 ⟨3.2%⟩	0 ⟨0.0%⟩	0 ⟨0.0%⟩	0 ⟨0.0%⟩	1 ⟨0.1%⟩	44 ⟨3.3%⟩
Public technology support organizations[4]		234 ⟨17.5%⟩	28 ⟨2.1%⟩	0 ⟨0.0%⟩	1 ⟨0.1%⟩	25 ⟨1.9%⟩	259 ⟨19.3%⟩
Higher education institutions		169 ⟨12.6%⟩	114 ⟨8.5%⟩	112 ⟨8.4%⟩	10 ⟨0.7%⟩	9 ⟨0.7%⟩	284 ⟨21.2%⟩
Other business firms within an establishment's own enterprise group		40 ⟨3.0%⟩	213 ⟨15.9%⟩	3 ⟨0.2%⟩	56 ⟨4.2%⟩	44 ⟨3.3%⟩	336 ⟨25.1%⟩
Business firms outside an establishment's own enterprise group		405 ⟨30.2%⟩	224 ⟨16.7%⟩	5 ⟨0.4%⟩	49 ⟨3.7%⟩	188 ⟨14.0%⟩	636 ⟨47.5%⟩
Private non-profit organisations and others		7 ⟨0.5%⟩	107 ⟨8.0%⟩	107 ⟨8.0%⟩	0 ⟨0.0%⟩	116 ⟨8.7%⟩	230 ⟨17.2%⟩
Overall		751 ⟨56.1%⟩	574 ⟨42.9%⟩	120 ⟨9.0%⟩	96 ⟨7.2%⟩	373 ⟨27.9%⟩	

Table A.7 (Continued)

Whether having cooperation arrangements on R&D activities with other organisations/Type of cooperation organisation[1]	No. of establishments having undertaken R&D activities	Location of cooperation organisation[1]					
			Mainland China and Macao				
		HK	Pearl River Delta (PRD) Economic Zone[2]	Pan-PRD Region[3] other than PRD Economic Zone and HK	Other regions	Places outside HK, the Mainland of China and Macao	Overall
Not having cooperation arrangements on R&D activities with other organisations	**3525** ⟨72.5%⟩						
Total	**4864** ⟨100.0%⟩						

Notes: (1) May select more than one organisation and/or location.
(2) The Pearl River Delta (PRD) Economic Zone covers an urban area of 14 cities and counties including Guangzhou, Shenzhen, Zhuhai, Foshan, Jiangmen, Dongguan, Zhongshan, Huizhou City, Huiyang county, Huidong county, Poluo county, Zhaoqing City, Gaoyao, and Sihui.
(3) The Pan-PRD region covers Fujian, Jiangxi, Hunan, Guangdong, Guangxi, Hainan, Sichuan, Guizhou, Yunnan as well as Hong Kong and Macao. Cooperation arrangements with PRD Economic Zone and Hong Kong are excluded in this column.
(4) Examples are Hong Kong Productivity Council (HKPC), Hong Kong Applied Science and Technology Research Institute Company Limited (ASTRI), Hong Kong Jockey Club Institute of Chinese Medicine Limited (HKJCICM), and R&D Centres.
Figures in square brackets represent the percentages to total no. of establishments having undertaken R&D activities.
Figures in round brackets represent the percentages to total no. of establishments having cooperation arrangements on R&D activities with other organisations.
Source: Adapted from HKCSD (2008b).

Notes

1. The investment period of the Applied Research Fund expired at the end of March 2005 and the Fund has ceased making new investments.
2. The six R&D Centres include an existing one, namely the Hong Kong Jockey Club Institute of Chinese Medicine Limited, plus five new ones.
3. This is quoted from an interview conducted by Adam Segal in June 2008.
4. External and overseas reviewers with international reputations are called upon to join the assessment committee.
5. See Douglas Fuller, Chapter 10 of this volume.
6. With particular reference to the Japanese version (MEXT, 2008), for example, in Article 7: "The government shall take the appropriate legislative, fiscal, financial and other necessary measures required to implement the policies with regard to the promotion of S&T"; in Article 9: "The Government shall establish a basic plan for the promotion of S&T in order to comprehensively and systematically implement policies with regard to the promotion of S&T." Such legal foundations may help Hong Kong to gather momentum in order to march towards becoming an "innovation hub" in the region.

References

Baark, E. and N. Sharif (2006) Hong Kong's innovation system in transition: Challenges of regional integration and promotion of high technology, in B. Lundvall, P. Intarakumnerd and J. Vang (eds), *Asia's Innovation System in Transition*, pp. 123–147, Edward Elgar, Chettenham, UK and Northampton, MA, USA.

Blue Book of China's Regional Development (2007) *The Development Report of China's Regional Economy (2006–2007)*, Social Sciences Academic Press (China), Beijing (in Chinese).

Chang, Yuan-Chieh, Meng-chun Liu and Phil Yang (2004) "The changing governance of academia innovation in Taiwan: A preliminary post-STBL review", *Taiwan Academy of Management Journal*, 4(3), 271–288.

Chen, Shin-Horng, Alice Lin, Pei-Ju Yu and Pam Wen (2008) "Open innovation and types of innovation: A policy perspective", powerpoint presentation at 2008 International Conference on Industrial Technology Innovation, Taipei, 21–22 August 2008.

Chesbrough, Henry (2003) "The logic of open innovation: Managing intellectual property", *California Management Review*, 45(3), 33–58.

Freeman, C. (1987) *Technology Policy and Economic Performance: Lessons from Japan*, Frances Pinter, London.

den Hertog, P. (2001) "Knowledge-intensive business services as co-producers of innovation", *International Journal of Innovation Management*, 4(4), 491–528.

Hills, Peter, Jacqueline Lam and Richard Welford (2004) "Business, environmental reform and technological innovation in Hong Kong", *Business Strategy and the Environment*, 13, 223–234.

HKCSD (2008a) "Statistics on research and development of Hong Kong, 2002 to 2006", *Hong Kong Monthly Digest of Statistics*, Feature Article.

HKCSD (2008b) *Report on 2007 Annual Survey of Innovation Activities in the Business Sector*, Census and Statistics Department, Hong Kong.
ITC (2009) Controlling Officer's Report.
ITC (2008) Innovation and Technology Fund Statistics of Approved Projects (as at 31/05/2008); URL:http://www.itf.gov.hk/eng/statistics/StatTable107View.asp?StatTypeId=107&StatId=500&StatCaption=Distribution+of+Approved+Projects+among+Different+Technology+Areas.
Lee, Woolgar (2006) *A Comparative Assessment of Training Course for Knowledge Transfer Professionals in the United States, United Kingdom, and Japan*, National Institute of Science and Technology Policy, Tokyo.
Liu, Wan-Hsin (2008) "Do active innovation policies matter?-Findings from a survey on the Hong Kong electronics SMEs", *Kiel Working Chapter No.1445*, Kiel Institute for the World Economy, Kiel.
Lundvall, B. (1992) (ed.) *National Systems of Innovation: Towards a Theorem of Innovation and Interactive Learning*, Pinter, London.
MEXT (2008) The Science and Technology Basic Law (Unofficial Translation), http://www.mext.go.jp/english/kagaku/scienc04.htm (accessed at 25/11/2008).
Nelson, R.R. (1993) *National Innovation Systems: A Comparative Analysis*, Oxford University Press, New York.
OECD (2005) *Governance of Innovation System, Volume 1: Synthesis Report*, OECD, Paris.
OECD (2008) *Science, Technology & Industry Outlook 2008*, OECD, Paris.
Sharif, Naubahar and Erik Baark (2008) "Hong Kong as an innovation hub for Southern China", http://www.hkjournal.org/PDF/2008_winter/6.pdf (accessed at 19/11/2008).
TechMatrix Research Centre (2008) *Innovated by Hong Kong*, TechMatrix Research Centre, Hong Kong.
Vonortas, Nicholas, S. (2008) "Innovation policy initiatives around the globe: A collection of good practices", http://www.gwu.edu/~cistp/events/Vonortas_Presentation_Innovation%20Policy_Good%20Practice_nv.pdf (powerpoint presentation, accessed at 20/08/2008).

6

Hong Kong's Venture Capital System and the Commercialization of New Technology*

Kevin Au
The Chinese University of Hong Kong

Steven White
China Europe International Business School

The objective of this chapter is to describe the features of Hong Kong's venture capital system and make policy recommendations to improve its effectiveness as an institutional support for the establishment of new firms and the commercialization of new technologies as part of a larger objective of diversifying Hong Kong's industrial base and creating a base for future economic growth. To do this effectively, we must address venture capital (VC) as a "system" rather than the more limited sense of a category of investment capital or a segment of the finance industry. With this broader scope, we will explicate the characteristics of specific types of actors and the formal and informal rules and norms by which they make decisions. As our analysis reveals, the current system has emerged from Hong Kong's particular historical, social, political and economic environment. While logical in this sense, it has not proven to be very supportive of new technology-based ventures.

Hong Kong has the largest pool of venture capital in Asia and has been home to one of the largest number of funds in Asia since the mid-1990s (Figure 6.1a, b). In spite of this apparent huge pool of investment funds, however, its performance in terms of financing new technology-based firms in Hong Kong has been low. In their World Bank study, Kenney et al. (2007) described the VC industries in Taiwan and Israel as successful, but those in Hong Kong as not. Indeed, we have found that the conclusion of the Massachusetts Institute of Technology (MIT) researchers in their 1997 government-funded study *Made by Hong Kong* (Berger and

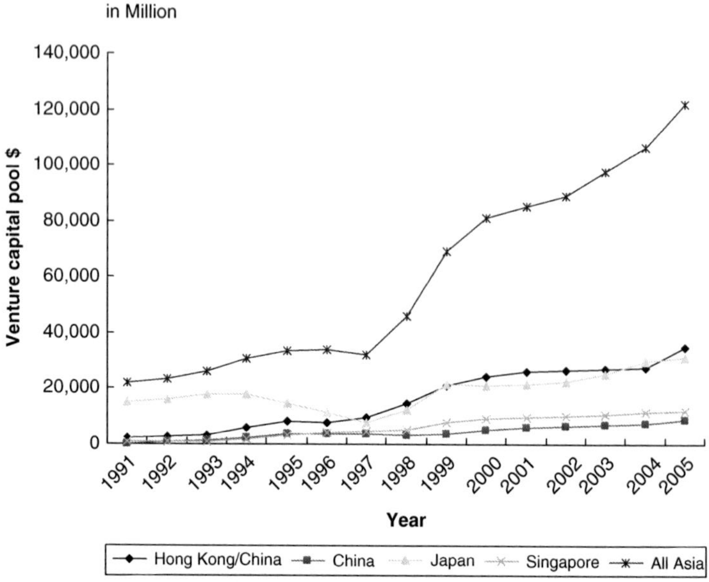

Figure 6.1a Venture capital pool, Hong Kong compared with other countries

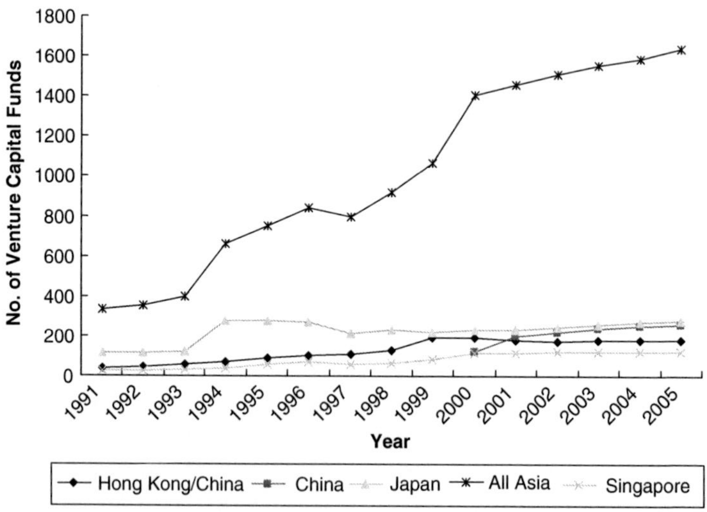

Figure 6.1b Number of VC firms, Hong Kong compared with other countries
Source: *Asian Venture Capital Journal*, 2000.

Lester, 1997) is just as apt today, more than 10 years later: "While there is plenty of capital available in Hong Kong, it is striking how little of it is directed into start-up firms or into funding technological upgrades. In addition, surprisingly few institutional investors have such investments as a focus of their strategy" (p. 293). While Hong Kong's finance industry as a whole has grown phenomenally since that time, much of this success is attributable to initial public offerings (IPOs) and investments at the expansion, mezzanine and buy-out stages, mostly in firms without a central technology focus.

The reasons for this situation can be traced to several features of the Hong Kong business environment; namely, a historically rooted trading and arbitrage business mentality, the legacy of British banking practices, manufacturers' reliance on short-term loans, and the particular backgrounds of locally active VC investors. These have combined to create a short-term investment mentality. In stark contrast to the so-called "classic" model that emerged around Silicon Valley in the USA, venture capitalists in Hong Kong avoid early-stage investments and seldom nurture early-stage ventures that are commercializing new technologies. Instead, they embrace a relatively short investment horizon and much lower risk threshold. Although such proclivity is common among venture capitalists in Asian countries (Kenny et al., 2007), Hong Kong's being a financial centre creates a subtle but powerful tendency for local venture capitalists to see investments as financial "pure plays" (Gupta, 2000). In this regard, they are more like the venture capitalists found in New York, who also take a more purely financial approach to VC investing compared with the classic model. The recent technology and internet ("dot-com") bust in 2001 and the resulting losses of many VC funds that had been invested in early-stage firms at that time have only exacerbated the situation. These factors together help explain why much fewer investments by Hong Kong funds are in early-stage ventures, especially when compared with Mainland China (Figure 6.2).

The irony is that in recent years, the local supply of potentially commercializable technology has been increasing. Hong Kong has begun to see the fruits of years of significant government support of university-based research and support in the form of incubation programs such as that in the Hong Kong Science Park. However, missing links and mismatched features of the institutional structure, investor cynicism towards technology investment, and a lack of mutual collaboration among key stakeholders (for example, between angel investors and the VC community) have obstructed the emergence of a new paradigm of

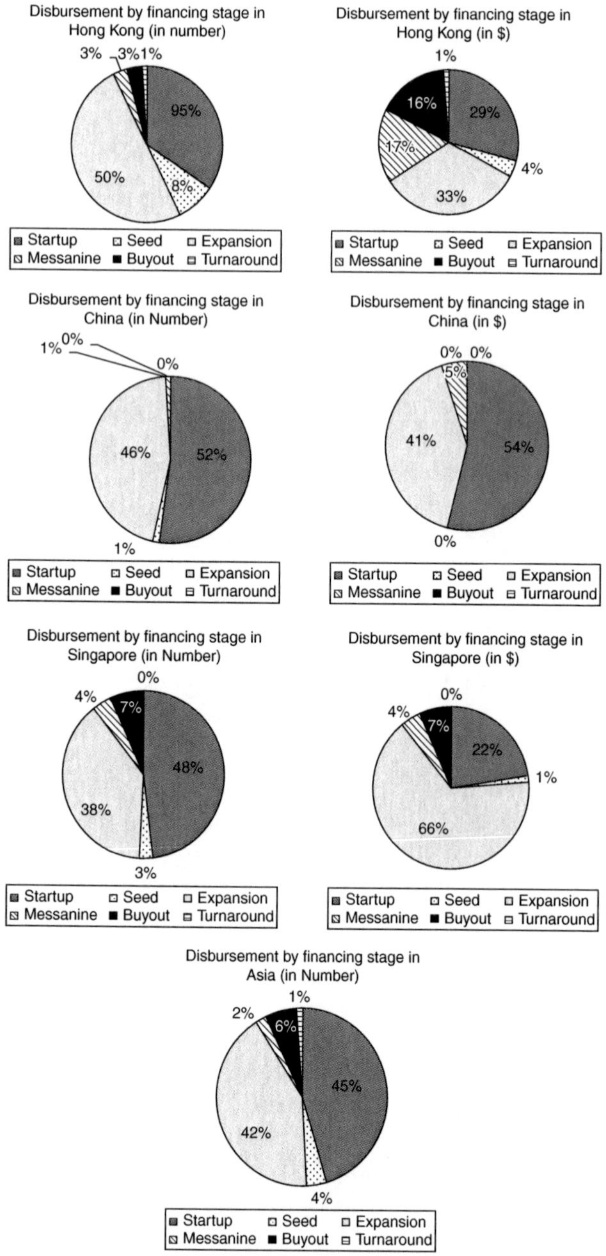

Figure 6.2 Investment stage of venture capital funds, Hong Kong compared with other countries

Source: *Asian Venture Capital Journal*, 2000.

technology-focused investing. The state of early-stage investment on technology start-ups, by so-called angel investors or informal venture capital (Mason, 2006), is a particularly weak link. The aftermath of these missing linkages may be summarized by the executive director of one international VC firm specializing in investments in technologies developed in universities, which opened its Hong Kong office in 2005, as follows:

> The society is innovative and entrepreneurial. The Science Park is great in innovation...But there is not a cohesive financial arrangement, and entrepreneurs, academics, politicians and civil servants lack the appropriate skills to differentiate good ventures from bad. They have to mature in their decision-making to accept risk and how return is generated.

In the rest of this chapter, we first place venture capital more formally within a broader institutional framework and use an evolutionary approach (e.g., Murmann, 2003; North, 1990; Whitley, 1992) to trace its emergence and key contextual features that have had an impact on its character. We can then understand the current status of investment in early-stage technology ventures. We go on to highlight the impact of the financial and industrial environment in general and the nature of angel investing in particular. This leads to a discussion of key factors that inhibit the function and performance of Hong Kong's VC system and our proposals for addressing these. In brief, they include policy recommendations to increase the funds available for smaller but higher-risk investments (i.e., to cover the "equity gap" of new ventures); to professionalize and diversify the skills of venture capitalists and private equity professionals; to professionalize and formalize angel investors and associations; and provide more direct support for nurturing new ventures.

1. Analytic framework

At the center of the venture capital system is the business of venture capital. This is typically defined as private capital that is pooled from investors and managed by professionals ("venture capitalists") who invest in seed- or early-stage new firms that have a potential for annualized gains of 30–40 percent or more to compensate for the high failure rate of others in the investment portfolio of the funds. VC firms usually realize the return on their investments ("exit") by selling the firm to another company ("trade sale") or after the firm is listed on the stock market (IPO).[1] Venture capital firms in the Silicon Valley made their names in nurturing what came to be prominent leading technology firms. In the book *Done Deals* (Gupta, 2000), based on interviews with founders of the venture capital industry, Larry Sonsini describes the scope of activities of these venture capitalists (p. 212):

> The providing of capital was one function of the venture capitalist. Being actively involved in developing the business model, managing the enterprise, and recruiting management... They thought of more than investing money. They thought about mentoring, training, and providing business solutions. The goal was not only to make a successful investment but also to be a part of building a successful venture.

As this business model became successful it attracted massive funds, and has since evolved to include opportunities for pure financial plays, such as financing leveraged buyouts and technology joint ventures. Because of this increasingly purely financial approach to venture capital, some would consider it to be a specific type of private equity (PE), which invests in private but not publicly traded companies (Metrick, 2007). Indeed, the use of private equity today resembles the traditional role that Wall Street financiers or English merchant banks played, using capital to organize and reorganize firms and industrial sectors. In Asia and Europe, the distinction between venture capital and private equity is less well defined than in the USA (Kenny et al., 2007). The name of the local industrial association, Hong Kong Venture Capital/ Private Equity Association, reflects this situation.

Because of sometimes important differences in terminology and practices, it is necessary to clarify what exactly is a "venture capital system" generically and then examine how it emerged and operates in a particular context. The approach of comparative business systems (Child, 2000; Murmann, 2003; Whitley, 1992), and as applied to the specific case of studying venture capital and private equity in different national contexts (e.g., Bartzokas & Mani, 2004; Kenny et al., 2007; White et al., 2005), provides a useful means of structuring such an analysis. Here we adapt the system-level framework of White et al. (2005) to the task of identifying key features and opportunities for improving the effectiveness and efficiency of Hong Kong's venture capital system (Figure 6.3). This framework has three features. First, the venture capital system

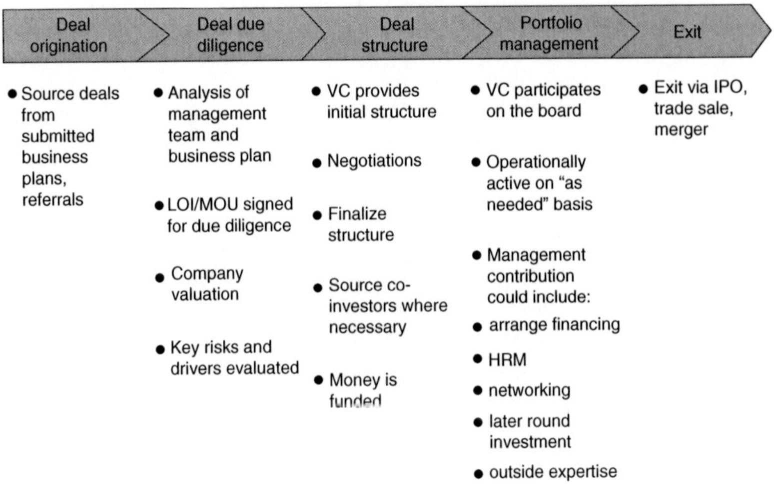

Figure 6.3 Key activities in the VC investment process

entails a fundamental set of activities that together support the commercialization of new technology. Research across diverse contexts (e.g., Amit et al., 1998; Black & Gilson, 1998; Jeng & Wells, 2000; White et al., 2005) suggests the following to be fundamental activities:

1. pooling funds
2. identifying investments
3. channeling funds
4. monitoring invested funds ("funds in-use"), and
5. appropriating returns to invested funds

Formal venture capital firms have institutionalized these into discrete steps (Figure 6.3), and these have become relatively standardized across national and regional contexts. As comparative research has found, however, there are important differences at a more micro-level of decision-making and investment management, which we will discuss further in the next sections focusing on Hong Kong's particular characteristics.

Second, the venture capital "system" is the configuration of institutionalized structures – actors and rules and processes – by which these fundamental activities are organized and integrated. The venture capital system, therefore, includes both the actors who undertake the focal activities of the system, as well as the regulations, practices and norms that have become established (or "institutionalized"). Relevant actors include angel investors, venture capital firms (both private as well as government-supported), and other types of investors. These actors have particular capabilities and preferences related to each of the focal activities. They also operate under formal regulations and informal industry norms.

Third, the system has emerged and evolved in response to Hong Kong's particular environment. Following White et al. (2005), we include in our analysis the relationship between Hong Kong's venture capital system and other institutionalized systems (education, legal, industry, etc.), as well as the material and ideational logics that have an impact on the system's structure and dynamics (Figure 6.4). Material logics are the economic or technological imperatives that also structure the choices of actors. These include the degree of competition,

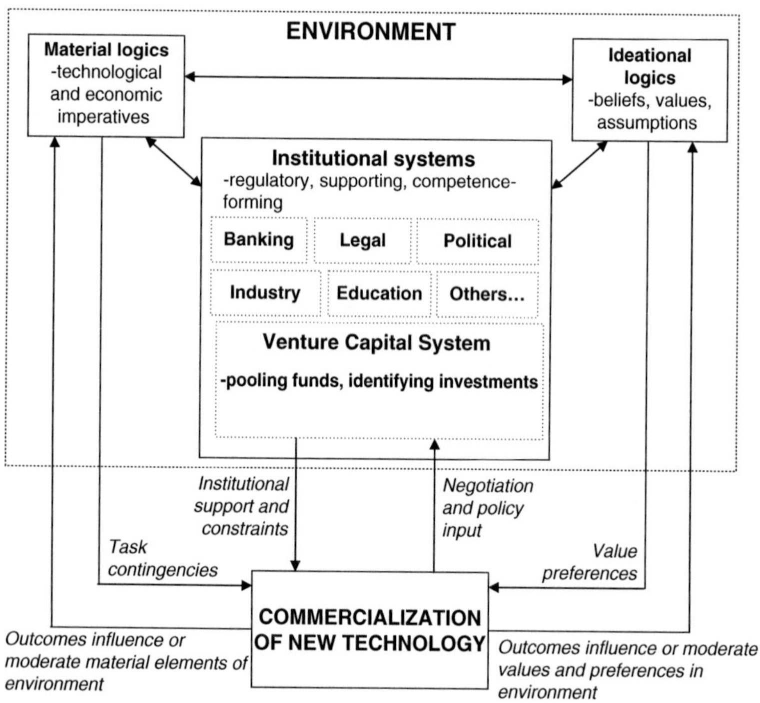

Figure 6.4 Venture capital system
Source: Adapted from Child (2000) and White, Gao & Zhang (2005).

level of human capital, interest rates, and so forth. Ideational logics are the beliefs, assumptions and values that define the preferences and influence the choices of actors.

2. Evolution of Hong Kong's venture capital system

In order to understand why Hong Kong's venture capital system has performed relatively poorly in terms of supporting the establishment of new technology-based firms, it is necessary to explore the historical context from which it emerged. To do this systematically, we apply the analytic framework introduced in the preceding section. We link the nature of the key actors, rules, practices and other features of the institutional environment to the impact on the establishment and support of new firms in Hong Kong.

2.1. Prior to the 1970s

Its geographic location, colonial development and control, and several historical events have combined to solidify Hong Kong's identity as a regional entrepôt with an economy heavily reliant on trading and related services. From the 1800s the British trading companies used the colony to extract resources from the Mainland. The Korean War and civil wars in China in the mid-1900s made trading a lucrative business when Hong Kong was a free port. Banks and trading houses flourished. Light industries also developed rapidly since refugees provided cheap labor for manufacturing, and more so as world trade expanded rapidly after World War II. This was further supported as Western countries began to outsource components and then completely transfer original equipment manufacturing (OEM) to Hong Kong.

The colonial administration did not implement sophisticated governance in this natural resource-poor territory; essentially, it relied on a system in which co-opted local compradors dealt with the locals. Bankers and more sophisticated industrialists who had fled from Shanghai, rather than local industrialists from Southern China, were preferred for this role. This approach was in line with Britain's fundamentally non-interventionist doctrine vis-à-vis the territory, along with its mandate to maintain a balanced budget. This structure remained fairly consistent over more than 100 years of colonial control, and it has left a major imprint on Hong Kong's society in general and administrative culture in particular. As a basic policy, the government did not take any explicit developmental role towards industrial development or any specific industry. Requests for direct subsidy for a company, an industry or a sector were routinely dismissed. In reality, public spending on infrastructure that supported Hong Kong's trading role – such as the construction of ports and an airport – did indirectly support the development of specific sectors, including trading, logistics, and construction. As a result, the Hong Kong environment was seen as being more supportive of these sectors than manufacturing, along with the finance and real estate development.[2]

2.2. 1970s to mid-1990s

Up to the 1970s, diversified trading houses (*hongs*), such as Swire and Jardine Matheson, and banks from the UK and a few other European countries dominated the Hong Kong financial scene. Even when the stock market soared, riding on the industrial and economic boom of the 1960s, large American financial houses, such as Merrill Lynch, served primarily US companies and individuals trading stocks on the US markets.

At that time, some Hong Kong-based companies with new products were able to take advantage of the frenzy and launch successful IPOs. Quite a few of these, however, had questionable products, failing after their stocks crashed along with the overall market with the onset of the 1972–1973 oil crisis.[3]

The local economy and the market, however, recovered quickly after the oil crisis. The British conglomerates and the emerging local tycoons continued to expand, executing huge deals such as Cheung Kong's purchase of Hutchison Whampoa from Hong Kong Bank. Hong Kong began to give rise to attractive investment opportunities in new businesses.

By 1972, Inter-Asia Venture Management had already been founded as the first formally organized VC firm in Southeast Asia. The three founders were classmates and had studied venture investment at Harvard Business School. R, one of the founders, recalled, "I saw my classmates and professors setting up funds, and I was determined to bring the venture investment model back when I returned to Hong Kong." As there was little technology in Asia, they invented a "transfer" strategy, in which they would focus on bringing proven advanced technologies (for example, photovoltaic solar cells) and service models (McDonald's & IKEA) from overseas. The first investors in their Inter-Asia Fund I were Sir Kenneth Fung and his family, in addition to other friends and families of the founders including Victor Fung.[4] Over 30 years since then, a similar group of investors has supported four of their funds.

The lucrative stock returns of the 1970s (averaging 25 percent p.a.) compared with the low return (several percentage points) that much of the public would accept for bonds, bank loans or savings accounts, created a huge profit gap and triggered more money going into what came to be called "direct equity investment." To take advantage of this gap, Arral & Partners was set up in October 1981 and would later rise to prominence.[5] Three of the five founders originally worked for US commercial banks in Hong Kong. They saw many opportunities, built up their networks through their work for the banks, and invested primarily their own personal funds. One of them, W, recalled, still with excitement, their pioneering approach:

> Both the local and international banking communities were skcptical. Most of the companies were not transparent, and building good relationships was the only way to get to know a company, so it's difficult to do due diligence or sit on the board to work with the management...Nonetheless, we proved them wrong, as other companies followed suit.

Arral was launched not particularly at the best time when interest rates were rising to historical heights. Yet it was able to invest T-Bills on a short-term basis at high rates (17 percent) while waiting to find longer-term investments. In 1982 Arral made its first direct investment in a company called Hong Kong Teakwood. This company later changed its name to Universal Furniture, and W cited it as the first Asia firm to do an IPO on NASDAQ (in 1984). This investment realized an internal rate of return (IRR) of about 45 percent. Following other positive investment outcomes, Arral raised US$30 million for its Arral Pacific Equity Trust I from pension funds, endowments, and other investors in 1988, and became well known to international investors seeking higher returns in Asia at that time.[6]

The year 1988 marked the launch of the largest fund in Asia at that time, the US$150 million APAC Fund, by a French bank called Suez (Asia). Its investor base was very globalized.[7] The person in charge, L, was a Hong Kong Chinese who had worked for a large accounting firm after his studies in the US. Seeing so many investment opportunities, L switched to a British investment bank which sent him, in 1981, to New York and Sand Hill Road to learn the business of venture capital. When he returned from his sojourn in the US, he moved to a French bank to head up what was called the Direct Investment Department because French banks had the practice of holding shares in companies that they helped list. He recalled:

> The financial atmosphere and practices were affected and shaped by the British at that time. There were only UK firms and brokers; those from the US did not have a real presence. British didn't like long-term investment. There was no medium-term money like bonds and debt instruments; short-term money was from bank loans and long-term money from equity.[8]

L claims to be the first banker to have supported Hong Kong factories, such as Johnson Electric and Playmate, which prospered with China's open-door policy in the 1980s. He realized returns of several times the US$10 invested in his fund.

Except for one early-stage wine venture in China, however, his investments were in late-stage investments (expansion and mezzanine). He noted, "One can make money from late-stage investments in 1 or 2 years and not have to run a company, which takes a lot of effort." He used his superb understanding of the economic cycle and regional development to find opportunities in Hong Kong (Tetronics, Instant-Dick), Taiwan (President Enterprise, D-Link), India (Zee TV), and ASEAN (Britainia Food, Q-Tel, Henley) during the business downturn of the late 1980s. He started selling in the early 1990s and returned all initial investments to APAC investors by 1997, realizing a return of 33 percent p.a.

By that time, many more funds had entered Hong Kong. AIG's Asia Infra-Structure Fund, for example, raised a staggering US$1.7 billion in 1994. APAC's success, the founding of Arral & Partners, and the establishment in 1987 of an industrial association, called the Hong Kong Venture Capital Association then, marked the emergence of VC as an investment form distinct from Hong Kong's traditional direct investments.[9]

2.3. Mid-1990s to 2001

Stimulated by the technology and economic boom of the 1990s, new VC firms popped up in quick succession. Some of them took a long-term view of investing in early-stage technology ventures. These included those set up by bankers and investors living in the region, for example Tech Pacific (later renamed as Crosby) in 1998, and those newly arrived from the Silicon Valley, such as Asia Tech.

The Asian Financial Crisis of 1997 was a setback to many investors who did not exit before the crash. The distress, however, attracted many large private equity firms to Hong Kong and Asia for the first time. Peter Brooke (founder of Advent International) said, "US institutions would not move into Asia [as venture capital]. After the Asian Crisis, many banks suffered, and the US capital that moved in was 'reorganization capital' (rather than expansion capital) from strategic investors, corporate investors, and multi-nationals" (Gupta, 2000). Newbridge, Carlyle Group, CVC Asia Pacific, and other large private equity firms set up their operations in Hong Kong as a base to look for distressed assets and buyout opportunities in the region.[10]

In 1998, with the effects of the Asian Financial Crisis cleared, and following the return of Hong Kong to China in 1997, the government set up the Innovation and Technology Fund (ITF) as a means of supporting the transition of local industries from labor-intensive and OEM manufacturing to higher value-added activities. Complementing this was the Applied Research Fund (ARF). It was a government-sponsored venture fund with HK$750 million in 1998 and about half invested by 2005. Its management was outsourced to three VC firms: HSBC Private Equity Asia, Walden International, and Asia Tech, with Tech Pacific later taking over Asia Tech's share. The ITF also set up the Small Enterprise Research Assistance Program (SERAP) in 2004 to finance R&D in start-ups, initially offering forgiving loans of up to HK$2 million or US$250,000, which has recently been raised to HK$4 million. Recipients are required to repay the loan only if their projects become profitable or they are bought out.

With the rise of the internet and dot-com bubble, many corporations in the "old economy" (infrastructure, real estate development, and logistics, in particular) tried to jump on the investment bandwagon. They set up corporate investment funds or entire subsidiaries to invest in technology and internet-related ventures. TOM.com and SunEvision, for example, were backed by large property companies and prominent business families. During this time, raising funds in the IPO market was very easy. While some of these established firms used the funds to develop new businesses that had synergies with their existing businesses, such as internet services for apartment dwellers, many of them also made eye-catching investments in unrelated start-ups. Similarly, if not more so than other local venture capital funds, they had a short investment horizon and did not aim at making money through nurturing new firms.

2.4. 2001 to present

After the dot-com bust, many new VC firms established in the late 1990s suffered huge losses and became dormant or were closed. The corporate VC funds withdrew from investing in the "new" economy and returned to projects in the familiar "old" economy. The internet subsidiaries, such as SunEvision and China.com, were quickly consolidated. Apart from the rapid economic downturn, there were several reasons behind their dismal investment records:

1. Many new ventures rode on the internet frenzy and actually had little new technology, lacked creativity, and produced only "me-too" products. Most venture capitalists interviewed suggested that there were few high-quality investment opportunities for them. To them, Hong Kong people seemed unable to think out-of-the-box.
2. The newcomers who joined the VC industry during the hype were accountants and bankers. They lacked operational experience. K was a venture capitalist who had retired from Walden International. He stated, "My firm recruited people, like myself, with operational and start-up experience. Local VC firms, however, preferred bankers, accountants or corporate financiers due to their common background." R of Inter-Asia traced this characteristic to a historical cause. "There was still a glass ceiling of some sort for Chinese who worked for [foreign] banks. Spinning off and raising funds for themselves was a way to gain autonomy. Investors would be less willing to trust their money to a Chinese entrepreneur who operates a factory than a banker who was seen as reliable and with good connections... Back then, raising a fund was just a way to get ahead. Everyone started talking about raising funds only in the past 10 or 15 years."
3. VC firms found that Hong Kong might have some good technology in its universities, but there was a lack of qualified managers and entrepreneurs who could commercialize technology. There were (and still are) many senior managers who were outstanding in their own work, but they avoided risk and did not understand technology.

An experienced venture capitalist with a technology background commented:

> VCs invest not in technology, but whether the team is able to build a business around the technology. Ideal teams must be intelligent in the sense that they can see complex [start-up] issues as a whole.

> They must also possess a high EQ because if they pursue a really large opportunity, they will need to overcome setbacks and frustration. Strong salesmanship is a character they also need in order to persuade people and get what they want.

This venture capitalist stated:

> Availability of high power experienced international management with P&L experiences, sales background and strategic marketing background are much needed to scale companies. I also see the OEM mentality still persist, and good tech sold cheaply... not trying to maximize the value. SME is also not interested to go up the value chain for fear of losing control. Many of the tech products are "me-too" with little differentiation nor innovation.

The ARF of the government also suffered, and its loss of more than HK$240 million drew public criticism. Apart from the above reasons which affected most funds, the outsourced funds under this scheme and the managing VC firms performed poorly because:

1. Governmental guidelines that might not make the best business sense tied the hands of the venture firms, including restrictions on investment targets and incentive structure.
2. The Fund's objectives and operating principles as a public policy tool (requiring transparency, accountability, and constant annual return) did not sit easily with the operating principles of VC investing which emphasizes risk-taking.
3. Government officials with little experience in venture investment scrutinized the investment process and requested multiple levels of approvals.

The firms returned the investment, and the Fund was dissolved in 2004. It also left a grudge between the industry and the government.[11] Public criticism of the losses reinforced the general belief that investing in new technology and start-ups was not suitable or attractive in Hong Kong. That perception remains strong even though several of the firms that received investments from the Fund – including Wise News, InfoTalk, and ecVision – are still operating.

Indeed, not all investments during this period were lost. The boom saw the founding of technology start-ups in computing (such as

Outblaze),[12] mobile services (e.g., Cherrypicks), video streaming (e.g., Vcast), internet applications (e.g., China.com), integrated circuit (IC) design (e.g., Solomon Systech), and other areas. They received funding from investors, survived the bust, and have since thrived. Solomon can be considered an outstanding investment success, with its 2004 IPO returning over HK$1 billion to investors and management. It has also drawn a host of foreign companies to the Hong Kong Science Park and has formed an IC design cluster there.[13]

SERAP should also be regarded as a success, despite its shortcomings which were similar to those of the ARF. Its formal mandate is to finance R&D, but in practice a number of technology start-ups in internet software (e.g., Editgrid and Radica), engineering solutions (e.g., Sengital), medical equipment (e.g., Colisa), computer gadgets (e.g., i-buddy), and others initially survived on the fund's seed money. A few early recipients, such as Dragonchip and Kanhan, and more recent recipients, such as Frenzoo, have received further funding from venture capital firms and other investors.

Other start-ups also survived the bust and obtained funding from other sources, and the technology cluster seems to be attracting foreign industry investors in addition to pure venture capital. For example, E-Crusade, an internet market service firm, was acquired by Razorfish (a subsidiary of Microsoft) in 2006 to extend that company's reach in China. The two founders were Hong Kong natives. One of them explained, "We came from multinationals, and our clients in Hong Kong were multinationals... Razorfish found us more trustworthy and easier to communicate with than a Mainland counterpart." A senior manager of the Science Park also claimed that Taiwanese firms have acquired several of their incubatees. The technology cluster seems to be attracting this type of foreign investment, in addition to pure venture capital.

The stock market rebound that began in 2003 opened the door for investors to exit some of the new ventures. Because of China's listing requirements, and capital and currency restrictions, it was not attractive for venture funds to exit their China-based investments in China's domestic stock market. Many of them found Hong Kong more attractive and easier to execute than NASDAQ, except for a few notable cases. IDG, for example, listed Kingdee on Hong Kong's Growth Enterprise Market (GEM) and TenCent on Hong Kong's main board. In this way, although Hong Kong had few local technology ventures to list, VC and private equity funds used Hong Kong as a base to realize returns from their investments in companies based in China and other Asian countries.

Top global funds, such as KKR, Oaktree Capital, and Bain Capital, began to set up their local Hong Kong offices from 2005. Recent successful listings of large Chinese corporations, such as Alibaba.com, and other smaller technology ventures, have further strengthened Hong Kong's place as a financial centre.

Many new private equity funds, such as Spring Capital and FountainVest, were also formed during this recent boom. Their founders tend to be experienced professionals who left large firms. As in the past, however, the new funds are mostly expansion funds and target non-local ventures, and most of them are focused on China. *Asian Private Equity 300* (2007, p. 45) noted, "Market observers are waiting to see how quickly dealmakers will invest their record-size funds... In greater Asia's still-underdeveloped marketplace, private equity is an unfamiliar form of financing, and the pace of transactions is slow."

As China continues to develop and expand, it presents both new opportunities and new threats to Hong Kong as a hub of venture financing. The banking systems, infrastructure, and stock markets all benefit as Hong Kong functions as a platform for activities in China. However, as technology, talents (especially Chinese returned from overseas), and opportunities all gather in China, it may be just a matter of time before the regulators, stock market, financiers, and small investors in China become mature enough to allow venture funds to do IPOs and exit their investment without Hong Kong's involvement. An insider of HKVC/PEA warned, "Many of the venture funds have already skipped Hong Kong and set up their offices in China... By the time its stock markets become mature and the RMB circulates more widely, Hong Kong as a base for venture funds will be lost forever." When that happens, Hong Kong may find even less financial support for technology start ups and innovation.

2.5. Angel investors and early-stage investment

Both entrepreneurs and the general public in Hong Kong have heard stories about VC investing in technology ventures. Yet many early-stage entrepreneurs mistakenly regard VC as the source that would bridge their capital needs between the seed and the start-up stages, usually a level around US$1 million. Venture capitalists, however, are very unlikely to invest such a small amount, and US$3 million is a commonly cited lower-threshold figure for them to consider. Since most Hong Kong VC firms focus primarily on expansion-stage projects, their investments are even larger.

As a result, many early-stage entrepreneurs cannot find funding to bridge the "equity gap" between what they can gather from personal sources initially and what they could solicit from venture capital funds. Not being able to find investors is even more ironic in Hong Kong because of the huge volume of capital that is based there. One Australian entrepreneur, who located his state-of-the-art internet application start-up in Hong Kong, lamented, "Hong Kong has the technology and the infrastructure, and is underrated as a place for start-ups ... but I did not know where to find investors." His first round of funding came from an Estonia-based venture capital firm with whom he connected by chance over the internet. Several of the authors of this book have also identified lack of early-stage investment, which is usually constituted by angel investors, as an obstacle for technology firms specializing in fields such as semiconductors and biotech.

Hong Kong has a significant number of informal investors, but they are not easily accessed by "strangers" and therefore are not sufficient to fill this gap. According to the Global Entrepreneurship Monitor, close to 8 percent of adults in 2007 invested in other people's businesses. This puts Hong Kong second among high-income nations. The problem, however, is that most of the businesses in which they invested were not technology-based ventures. The investments were very small (averaging US$100,000), and belonged to friends or relatives. Thus, although the numbers suggest capital is available and there is some willingness to invest, only a very small fraction of these informal investors (less than 1 percent of the adult population) are "angels" who would invest in a "stranger's" venture.[14]

We interviewed self-described angel investors, and found that they fall into five categories:

1. ***Sophisticated***: They have entrepreneurial experience or professional backgrounds and manage a portfolio of ventures using the formal US model of angel investing. Some work with other angels informally, and some even form themselves into syndicates.
2. ***Businessmen***: They are either working or retired businessmen or professionals who invest in start-ups as an alternative investment form. Their daily business allows them to find investment opportunities. They generally follow the angel investment approach, but are less professional and operate more informally.
3. ***Corporate***: They are manufacturers or other types of firms who look for technology start-ups to extend their existing product lines or services. They may invest in kind (such as providing lab and engineering time) instead of cash.

4. ***Incidental***: They are well-to-do individuals who invest out of interest, as a challenge to prove themselves (for gaining face among peers), or with the desire to kill time. They invest conservatively.
5. ***Traditional entrepreneurs***: They are businessmen who carry the traditional Chinese entrepreneurial mindset of distrusting outsiders and requesting control. They prefer control and a majority ownership. Their approach is simply an extension of their business approach. They are either unaware of or purposefully disregard modern angel investment practices in other developed countries.

Among these, the last two types of angels are unattractive to technology start-ups, unless they are exposed to more sophisticated angel investment techniques. Cooperation between them and entrepreneurs is usually difficult, as such angels leave insufficient incentives and autonomy to entrepreneurs.

While the first three types do exist in Hong Kong, they are not very visible. Unlike their counterparts in the US and the UK, few organize themselves into angel syndicates. The government has yet to recognize this feature as a weakness in the local venture capital system. Several organizations do exist but there is a lot of room for improvement. Monte Jade and Hong Kong Angel Capital Network claim to have many members, but have yet to install collaborative procedures regarding deal flows, due diligence, shared investment or exit for their members. Other organizations involve only a small number of angels; Dark Horse Investment (formerly Asia Angel Association), for example, is composed of only six members with start-up, VC, operation, and management experience. They see the recent economic downturn as presenting them with particularly good opportunities. Two members noted during our interviews, "Calls have dramatically increased in the past few months... Valuations have gone down and are closer to our own numbers."

Despite the number of informal investors, Hong Kong still has an "equity gap" in providing initial funds to technology ventures for several reasons. First, angels fail to organize, probably because they simply do not see it as the best investment of their own time. In comparison, the Singaporean government has established the Business Angel Fund Co-Investment Scheme and Startup Enterprise Development Scheme to promote angel investments.[15] Second, it is difficult for entrepreneurs to locate and then pitch to angels. Like angels elsewhere, Hong Kong angels do not solicit proposals from strangers, remain low-key, and limit themselves to their own network of familiar faces. They act on referrals from reliable sources, which helps to reduce risk since the management team is the primary criterion for an early-stage investment. Finally, often angels may not be situated in Hong Kong or may prefer not to invest locally. Unlike angel investors in the USA who, typically, confine their investments to a geographic radius of a 3-hour drive, Hong Kong angels, like Hong Kong venture capitalists, prefer ventures in the Mainland rather than those that are Hong Kong-based, even if visiting and monitoring them takes much more time. As a chain is as strong as its weakest link, lacking early-stage, angel investors is something Hong Kong must address in order to boost more VC investment in technology start-ups.

3. Current structure: Analysis of key actors

This section describes each of the key actors who currently play important roles in the financing of the commercialization of new technology in Hong Kong. The interrelationships and flows of resources among them are depicted in Figure 6.5.

3.1. Government funds and agencies

The ITF was established in 1998 to provide funding to support technology start-ups and innovations in established corporations. We have already described the different degrees of success of the ARF and SERAP,

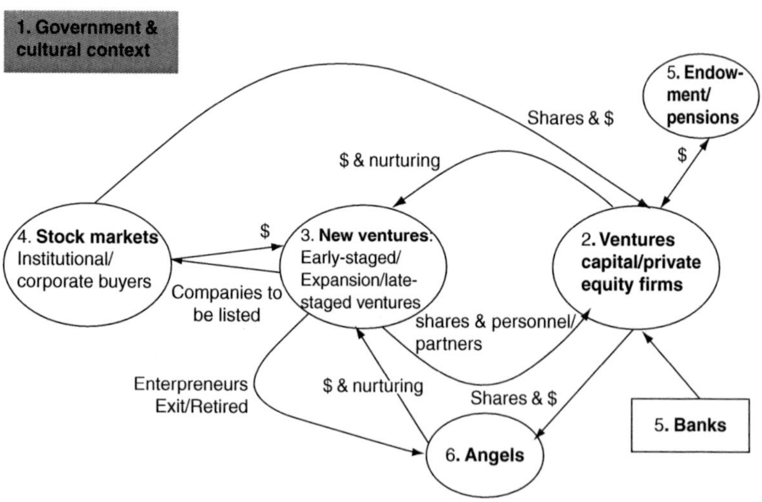

Figure 6.5 Key actors and resource flows in Hong Kong's venture capital system

two of its initiatives. In addition, governmental incubators, such as the Cyberport and Hong Kong Science Park, also subsidized their incubatees on rent, training, and marketing.[16]

The other major funding that has relevance for new technology ventures is the Small and Medium Enterprise Fund. This was set up in 2001 to improve the financing and competitiveness of SMEs. The funding schemes are not specifically targeted at new technology firms or the commercialization of technology. The Fund is administered by the Trade and Industry Department rather than the Innovation and Technology Commission (ITC) and, in practice, caters to the needs of existing SMEs for the purchase of equipment (by guaranteed loans) and by providing working capital, marketing, and training. Most SMEs have drawn on the Fund's allowance of up to HK$8 million to purchase advanced equipment in order to improve their operational capabilities and performance. Start-ups, however, would rarely be qualified to use the Fund.

3.2. Banks

Hong Kong's banking system continues to be among the most efficient in the world and is quite effective at supporting the expansion of established corporations. The recent arrival of large institutional investors has certainly broadened the business scope of Hong Kong's banks and opened up new opportunities. Most of them have diversified from their core business of mortgages and syndicated loans into wealth management and private banking. In addition, the establishment of individual credit ratings has also allowed many of them to step up their consumer financing services. Geographically, most of the larger local banks have entered China to tap the growing market. At the same time, more banks from China and overseas have set up branches in Hong Kong, and acquiring smaller family-owned banks has been one way that foreign banks have secured a beachhead in Hong Kong's vibrant, if increasingly competitive, financial market.[17]

These industry developments, however, have not changed anything with regard to the "brick" culture of local bankers. They remain quite traditional in their business practices. They almost always demand collateral (usually properties, letters of credit, or company shares) to support a loan for business investments. To lend working capital to a company, they will require the company to be profitable, and they will expect to examine 2–3 years of company records. On top of this, they almost always request a personal guarantee, usually by shareholders.

Even if the loan is guaranteed by the government SME Fund, the approval procedures and requirements are similar. Lending based on just a sound business plan is almost unheard of.

3.3. Institutional investors: Pension funds and university endowments

Pension funds in Hong Kong have grown to a significant size. According to the Mandatory Provident Fund (MPF) Schemes Authority, the government-mandated MPF had grown to over HK$220 billion in 2008 since its establishment in 2000. Other registered retirement schemes (OROS) totaled more than HK$250 billion during the same period. Their permissible investments are publicly traded equities, debts, warrants, and futures. Figure 6.6 shows the investment sources of VC funds in Hong Kong, China, and Singapore. In the USA, financial institutions provide only 18 percent of capital, while endowments and foundations provide 17 percent (Metrick, 2007, p. 27). In Asia, most funding is from corporations and insurance companies, followed by banks. Endowments and foundations do not play a significant role.

The US case suggests that university endowments could and should have more leeway in their choice of assets.[18] However, because Hong Kong universities are publicly funded, government guidelines restrict such funds to be invested in highly rated asset classes (although warrants and options are inherently risky), and not in hedge funds and private equity funds.[19] Although funds that originally came to the endowment from private donations and programs for foreign students are not public money, these began to trickle in only after educational policies changed several years ago. The total amount is still small. Furthermore, although recent investments have diversified, the general public is not yet willing to accept the possibility of losses in the case of endowments, according to a senior financial manager of a local university. He stated: "The executives from large corporations who sit on the board [overseeing the investments] do not want to be held accountable for losses…Changes may only be possible with the entry of more private universities, as new practices may then be introduced."

3.4. Stock markets (Main board and growth enterprise market)

Stock markets in Asia-Pacific are considered to have higher systemic risk due to the nature of their national financial systems. Even so, they have enjoyed a significant growth in IPOs and above-normal returns

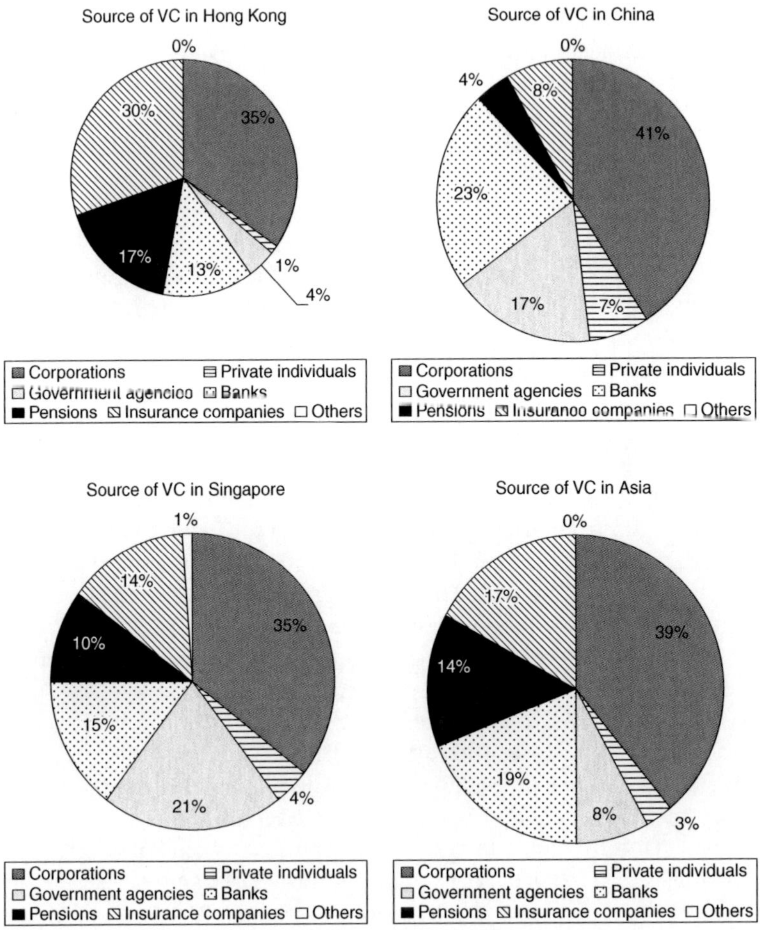

Figure 6.6 Sources of venture captial funds, Hong Kong compared to other countries
Source: Asian Venture Capital Journal, Year 2001.

because new issues are often underpriced. Although new issues would be overpriced during stock frenzies, that actually gives an even greater incentive for venture capitalists to push their investees to go IPO.

Also, compared with IPOs in other markets, the original owners and founders of newly listed firms in Hong Kong and other Asian markets retain a relatively high level of ownership and continue as management. In many cases, the objective of the IPO is not to sell out, nor to raise funds for growth, but for the firm to gain reputation. It also helps them bring in outsiders to improve management and governance.[20]

Riding on the growth in China in particular, and in the region in general, the Hong Kong Stock Exchange (HKSE) became the sixth largest stock market in 2006 (total market capitalization of over HK$22 trillion, or US$2.84 trillion) when it raised more funds than any other markets except London. A 2007 study commissioned by the City of London rated Hong Kong third, behind only London and New York, as a financial centre. It also picked Hong Kong as the most likely Asian candidate to develop into a global financial centre. HKSE, indeed, raised more funds in IPOs than any other market in 2009.

This spectacular transformation is largely attributable to the listing of Chinese corporations registered in China (H-shares) and the Hong Kong-incorporated subsidiaries of Chinese firms (including the so-called *red chip* firms with heavy influence from their Chinese parent companies). The listing of the first two red chips, Chinese enterprises CITIC Pacific and Guangdong Investment, at the beginning of the 1990s, was followed by a slew of red chip IPOs. Their success, in turn, stimulated the H-share IPOs, starting with Tsingdao Beer in 1993. An increasing number of regional governments restructured their industrial holdings and packaged their assets into "window companies" for listing. Excitement and speculation surrounding these listings reached a peak in 1997 and crashed with the Asian Financial Crisis.

In spite of that crash, the listing of China Mobile in the same year signaled the arrival of gigantic state-owned enterprises under central government control, and with it a new stage in the market's development. One by one, China's central ministries would restructure its industrial assets spread across multiple provinces and package them together to be listed as a whole new company. This was linked to China's broader economic and industrial structural reform. Following the telecommunications companies, those in petroleum, insurance, coal, chemicals, and other industries followed. By 2007, with the listing of two gigantic banks and the railway corporations, the IPOs of these central state-owned enterprises were coming to an end. As of February 2008, the 106 H-shares listed in HKSE had a total market capitalization of HK$4.4 trillion (US$568 billion) and accounted for close to 50 percent of daily transaction volume. Future listings of such firms, however, will only be co-listings in Hong Kong and China (so-called "A+H shares") since China's stock markets have matured.

As Chinese enterprises have increased the size, variety, and depth of their financial activity in Hong Kong, this in turn has drawn more institutional investors, hedge funds, private equity funds, and private banks to Hong Kong. Their presence has stimulated the need for new services, financial innovations, and sophisticated financial products. One result is that Hong Kong is now the most sophisticated derivatives market in Asia.

Another positive feedback from the large and sophisticated cluster of financial institutions has been to entice other types of companies to tap the Hong Kong market. Taiwanese corporations, for example, have begun to list their China assets in Hong Kong. Even more important has been the inflow of private Chinese enterprise listings, many of which were invested and nurtured by venture capital and private equity firms based in Hong Kong. Many of these are technology-intensive companies, such as BYD, Alibaba, and TenCent. For these listings, however, HKSE is competing with several other markets, such as NASDAQ, the London Stock Exchange's Alternative Investment Market, and the Singapore Stock Exchange. Working against HKSE is the fact that new technology firms are not the favorite type of firm for most Hong Kong investors.

The Growth Enterprise Market (GEM) was opened in November 1999 with the intention of addressing this situation. Its stated objective was "to provide capital formation for emerging companies to facilitate their business development and expansion," and it was designed for firms with high growth potential but which were unable to meet the track record requirements of the main board. GEM was established with the clear principles of "buyers beware" and "let the market decide." Although GEM had moderate success before the dot-com bust (raising HK$45 billion in the first 7 years), it now fails to attract technology firms, has lost much of its capitalization, and suffers from low liquidity. As shown in Figure 6.7, most VC/PE backed IPOs (between 2000 and 2007) listed on the main board are non-technology firms. The GEM has more technology firms, but most of them are actually based in the Mainland. It had to wait until 2009 before seeing another acclaimed local IPO.[21]

In 2008, HKSE decided to reposition GEM as a second board, and this has made it even less relevant for start-ups commercializing new technology. New GEM applicants are required to have a positive cash flow of not less than HK$20 million in aggregate for the two preceding financial years. Other procedures concerning the approval of applications and the transfer to the main board have been streamlined.[22] HKSE stated that its ability to list sound companies is what attracts investors, especially those from China, and regarded insufficient institutional investors, lack

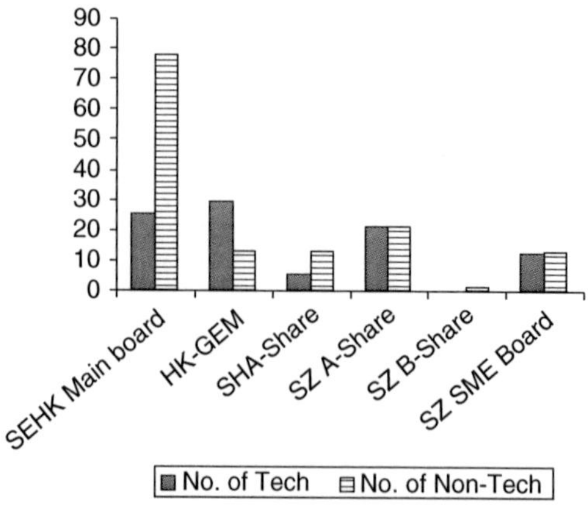

Figure 6.7 VC-backed initial public offerings, 2000–2007
Source: *Asian Venture Capital Journal.*

of tax breaks, and immaturity of investors as hurdles against a local market similar to the Alternative Investment Market of the London Stock Exchange (i.e., one that favors professionals as investors, higher risk, earlier-stage firms, and listings).

Accordingly, this transformation has attracted criticism, especially from the venture capital and private equity industry. An insider in the HKVC/PEA sees the new GEM as reflecting a critical difference in interests: "Less regulation will encourage people to take risk... HKSE has their eyes on attracting large international corporations and funds [which are the mainstay of HKSE], but that is probably not good for start-ups and VC firms."

4. System features inhibiting the commercialization of new technology in Hong Kong

Our review of the historical roots and evolution of Hong Kong's venture capital system and related developments in the financial and industrial sectors has uncovered a number of factors that inhibit the effectiveness of VC in supporting the commercialization of new technology in Hong Kong. This section summarizes those inhibitors which, as illustrated in Figure 6.5, should be conceived as an institutional system that unfortunately works against commercialization of technology start-ups in Hong Kong.

4.1. Governmental and cultural context

Hong Kong's business culture is strongly tilted towards a short-term investment mentality and a focus on low-tech industries and services. Government policies have been made with banks and traders in mind, and over the years this has helped Hong Kong grow into an international financial centre. Further effective lobbying by the banking and financial sector has led the government to continue to channel more resources to these sectors and further embed the short-term, quick-return attitude across society. Longer-term investments in R&D and risky innovation are selected against by a doubtful, cynical attitude towards start-ups attempting to develop and commercialize significantly new technology.

On the other hand, politics and ideology must also be considered. Changes progress fitfully as society and the government continue to debate as to whether Hong Kong can deviate from the positive non-interventionism doctrine. So when the government steps in, though its intention is to maintain fairness and accountability, micro-management and monitoring practices derived from a trader's mentality actually causes more trouble. Any governmental attempts to take Hong Kong on a different tack are doomed to fail (Goodstadt, 2005).

4.2. Venture capital and private equity firms

Most venture capitalists have accounting or finance backgrounds and adopt the attitude that "they are out to make money" rather than to nurture technology and new ventures that could have a major impact on an industry or even on the world. For them, the easiest way is the best way to make money, and this leads VC and PE firms to favor late-stage, mezzanine, and buyout deals. These deals are more widely available in Hong Kong (rarely are there good technology start-ups) and less risky while larger in size, and do not use up as much their time compared with early-stage deals. Indeed, few venture capitalists have the skills to build a company from scratch because they do not have the relevant experience themselves. Furthermore, their finance background biases them to hire others like themselves rather than former entrepreneurs and operational professionals.

4.3. New ventures

Many of Hong Kong's new firms are family businesses. These first-generation entrepreneurs are happy with the local market and business practices. They do not like transparency, guard their ownership very carefully, and tend to use insiders rather than professional managers (who were not available in the past). These characteristics are found broadly within society, and make it difficult for venture capitalists to invest in and work with these firms.

Some second-generation leaders of family businesses and the new generation of entrepreneurs are different. They have a world view and want to expand beyond the local environment. They are more receptive to transparency and are more likely to be able to introduce outside professionals and capital. However, a small local market, high costs, lack of advanced technology, scarce start-up capital (see below for a more in-depth discussion on angel funding and SBIC-like companies),[23] and a shortage of capable entrepreneurial teams make technology ventures more difficult to establish than non-tech ventures. In sum, although Hong Kong is famous for its entrepreneurial spirit, high-quality technology start-ups are rare. There is even a more general concern that Hong Kong's legendary entrepreneurial spirit is weakening and has been fading away.

4.4. Stock markets

The Hong Kong Stock Exchange welcomes the listing of large corporations (especially those from the Mainland) and has introduced more advanced financial products. These are the basis of its profits and attract large institutional investors, like pension funds and investment banks. The technology level of potential listees is not an important criterion for them.

4.5. Banks and institutional funds (Pensions, endowments)

Banks have a strong bias towards lending based on collateral rather than on the soundness of a business idea or the competency of a management team. They do not have the ability to assess and are quite reluctant to provide finance for start-ups and early-stage firms.

Retirement and endowment funds may invest as limited partners in VC or PE funds. Although their investment horizon and objectives would seem to be in line with the classic VC model that nurtures early-stage ventures over a medium- to long-term horizon, their impact in Hong Kong has been minimal due to Hong Kong's version of the "prudent man" rule.[24] As a result, money goes into and breeds ever larger and more expansive buyout funds. To nurture more technology start-up, the abundant capital "parked" in Hong Kong needs redirection.

4.6. Angels and Angel groups

The traditional way to finance new ventures is by savings and family capital, in addition to partnerships with friends and co-workers. Those in Hong Kong, however, tend not to organize themselves into syndicates, and there is a lack of institutional support and understanding of angel investing. For the small group of cashed-out or retired local technology entrepreneurs, there are few companions for them to co-invest with in Hong Kong and instead they focus their efforts on the Mainland. As a result, they do not fill the "equity gap" faced by technology entrepreneurs in Hong Kong.

5. Options for improving the performance of Hong Kong's venture capital system

Bearing in mind the limitations in overhauling fundamental institutional structures and beliefs, such as the willingness and legitimacy of the government to make major ideological changes, high costs, cynicism towards investment in new technology, and a short-term orientation among society, we propose the following policy options for implementing in the medium term. Due to the interrelatedness of these elements, they would be more effective if implemented together in a concerted effort.

1. Stimulate more VC funds with a longer time horizon and greater focus on new technology commercialization. One way is to encourage long-term investors to become limited partners of VC funds.
    - The government should channel university endowment funds and other government funds to VC funds since their longer time horizons are compatible. This should also attract more foreign funds and, at the same time, reinforce Hong Kong's position as a financial hub. One option would be to invest a small amount of governmental reserves (say 0.05 percent) as a legitimizing gesture and thereby encourage the endowment funds (estimated to be over HK$50 billion) to follow suit (say 0.5 percent).
    - The implicit "prudent man" rule could be lifted for MPF and other retirement funds, allowing them to invest a portion of such funds in non-publicly traded investment funds. PE funds may be more suitable given the investment objectives of retirement funds. A mere 1–2 percent of these funds (total HK$600 billion) would

amount to over HK$10 billion and bolster the outlook of many local VC/PE funds, instrumental for keeping investment talent in Hong Kong.

- ARF was reviewed and many lessons learnt after its closure 5 years ago. Israel's Inbal program was a failure but lessons learned led to its hugely successful Yozma program. Politics aside, the government may learn from these experiences and sponsor investors to form new VC funds that focus on technology start-ups. Such a new program should incorporate new features as informed by the Yozma program (for details, see Avnimelech & Teubal, 2004):

    - employing technology experts to administer the funds;
    - giving full autonomy to the VC companies on investment decisions; and
    - providing strong incentives on the "upside" for the funded companies (i.e., the possibility, within a median period, of purchasing government's share at cost), but no downside "guarantee" of losses.

2. Develop professional training and qualifications for investment advisors in the VC and PE industry, angel funds, and private companies.

    - In the US, investment advisors appeared in the mid-1980s to advise institutional investors about venture investments after the US "prudent man" rule was lifted (Gompers & Lerner, 2000, pp. 8–10). They pooled resources from their clients, monitored existing investments, and evaluated new funds. They helped stimulate the growth of the VC and technology industries in the US. Hong Kong can establish qualifications similar to the Chartered Financial Analyst (CFA) and Certified Financial Planner (CFP) qualifications in order to facilitate the investments of endowment and pension funds (complementing our recommendations in Point 1 above).

3. Stimulate angel investments.

    - Expand and professionalize angel investment by developing guidelines, successful/failure case studies, and templates for documents such as term sheets.[25] Commission HKVC/PEA and universities to develop related training courses and networking events.

- The Enterprise Investment Scheme (EIS) of the UK can provide several policy options for Hong Kong. The UK experience has been reviewed more and seems to work well (Mason, 2006).
- To establish the status of "Accredited Investors" to define "high net-worth" individuals.[26] They form the basis of angel groups because the status gives credibility to other angels and entrepreneurs. Also, it ensures that only informed investors are involved in risky angel investment. The proven system in the UK should be implemented in Hong Kong. Basically, individuals can self-certify to the government if (1) they have high earnings (say HK$1.5 million) or own valuable net assets (say HK$5 million), and (2) possess experience as a sophisticated investor in private companies, such as being an experienced member of an angel network, sitting on the board of or serving as professionals for private companies, and having experience in running or investing in these companies. Presenting false statements is against the law.
- Some of the tax incentives EIS used to encourage angels to invest in private companies can be adopted even though Hong Kong has no capital gains tax. These include a tax relief at the basic rate and income tax relief on losses. The relief rate can be defined after careful study. Investors can invest up to HK$2 million per annum and must hold on to the investments for at least 2 years. Perhaps higher tax breaks could be given to accredited investors who risk their money to invest in new, technology firms. Besides, such breaks should apply not just to investments in Hong Kong, but also to those in Shenzhen which has another significant presence of technology and people. The cities are sister cities and in light of more integration initiated by the central government,[27] more angel investments across the border would benefit Hong Kong in the long run.

4. Establish Small Business Investment Companies (SBIC) types of program to stimulate investments in small technology businesses and help to fill the equity gap.

   - The Enterprise Capital Fund of the UK was modeled on the US experience and has been viewed as successful since 2002 (Mason, 2006). A similar program should be implemented in Hong Kong. In essence the government would solicit competitive bids from qualified individuals (or companies) for plans to invest in small private companies (range HK$1 million to HK$15 million). The government would match up to twice the

amount raised by the bidder to form a fund, but would take a smaller share of the profits and an equal share of the losses. The investment period must be longer than 2 years. More careful study could define the maximum size of the government's match fund, the business nature, and technology content of the invested companies.
- The Start-up Enterprise Development Scheme (SEEDS) and Venture Investment Support for Start-ups (VISS) programs in Singapore require direct government involvement in making and monitoring decisions (Wang, 2004). They may not be as feasible in Hong Kong due to different administrative tradition and lack of qualified people in the government.

5. Diversify the backgrounds of the general partners of VC firms.

- Encourage the VC/PE firms to recruit special partners who are retired or cashed-out entrepreneurs to complement the jobs of general partners who tend to have an accounting or financial background. If advisors are more widely available, governmental and endowment funds may be stipulated to be invested only in firms with partners fitting this profile.
- With more active and organized angel groups, the inter-flow of talents between VC firms and other actors in the venture capital system should also be improved, and that should help diversify the backgrounds of the general partners.
- Target the recruitment of experienced general partners or entrepreneurs from the large population of Hong Kong expatriates living in global centers of innovation, such as Silicon Valley. Be sure to include them in a new reformatted ARF proposal.

6. Create new IPO exit routes for VC and PE invested in technology firms.

- It's almost a lost cause to change the HKSE which monopolizes company listings and stock trading.[28] One possible future exit lies in setting up a new board with perhaps the main board or the SME board of the Shenzhen Stock Exchange. Otherwise, IPOs in other markets like NASDAQ and other exit routes will continue to be the preferred choices. An efficient exit route would encourage VC investment in the long run and help Hong Kong to maintain its position as a financial centre.

Notes

* We would like to thank *Asian Venture Capital Journal* (AVCJ) for providing relevant data, and Boaz Wong and Jody Wong for their assistance in data analysis. The preparation of this chapter is partially supported by a grant from the CUHK Li & Fung Centre for Supply Chain Management and Logistics and a GRF grant of the Hong Kong University Grants Committee (CUHK 440008) given to the first author.

1. Trade sales usually have much lower returns than an IPO, and so are considered a second-best exit by the venture capitalists.
2. See Goodstadt (2005) and Au et al. (2005).
3. Hong Kong Antenna is one highly publicized example.
4. Victor Fung is a prominent figure behind the development of the VC industry, and was the founding chairman of HKVCA in 1987.
5. Although Transpac was started in the late 1970s, it originated from Singapore. Pica was also around in Japan in the early 1970s according to R.
6. Arral & Partners raised US$150 million on their APT II in 1991 (US$25 million from IBM's pension funds). Yet the partners split later in 1993 (Sender, 1993).
7. Including Princeton University Endowment, Loyola University Endowment, insurance companies from Japan and Western Europe, Ronny Chan's family fund, one Taiwanese family fund, and the government investment funds of Saudi Arabia and Singapore.
8. According to L, three British merchant banks at that time dominated the scene: Schroders, Jardine Fleming, and Wardley, which subsequently merged with Hong Kong Bank. Their main business was to give advice on IPOs or M&A. Quite a large number of British and local commercial banks were also around. Following the British banking tradition, their business was to service clients on loans. They rarely held equities and were usually short-term oriented, meaning the duration of their loans was shorter than an economic cycle. Among them, Hong Kong Bank might be an exception because its large saving deposits enabled it to sit through an economic cycle for some high-prospect clients without calling back their loans.
9. AIG Investment Corporation (Asia) Ltd and Prudential Asset Management Asia (PAMA) were especially active in direct investments between the mid-1980s and the end of the 1990s.
10. See Gupta (2000) and Cheng (2004).
11. Yet, it does not mean that government-sponsored programs all failed. Israel's program called Yozma did away with government bureaucracies and brought about the right incentives, compared with a previously failed program called Inbal. The Yozma program created a highly successful VC industry outside the US (Avnimelech & Teubal, 2004).
12. IBM acquired Outblaze in 2009 to gain access to its technology. The Cyberport has housed Outblaze during its development (Retrieved 17 December 2009 from http://www.bilal.ca/ibm-acquisition-of-outblaze).
13. InfoTalk has been sold but continues to exist as an independent company. Solomon Systech was a spin-off from Motorola with customer and revenue

support from the Taiwanese capital. So its success may not be considered on the same par as other Hong Kong start-ups.

14. Even in the recession years of 2001–2003, the angel investment rate is around 3 percent. See *Global Entrepreneurship Monitor 2007 Hong Kong Report* (www.cuhk.edu.hk/centre/entrepreneurship).
15. See Koh (2006).
16. They subsidized internships, promotion costs, professional fess, and rental of equipment in the range of US$600,000 in 3 years during the incubation.
17. China Merchants Bank Co. Ltd acquired a majority stake in Wing Lung Bank (Retrieved 21 December 2009 from http://english.cmbchina.com/CMB+Info/news/CMBnews/cmbnews2008060201.htm).
18. Morton Collins in the book *Done Deals* commented, "The investment objective of pension funds is not compatible with the high-risk, high-reward, early-stage, long-term, high-technology investing of the VCs. Pension funds have a target annual IRR, and are anxious to receive distributions of cash and securities." University endowment funds may have more compatible investment goals. "They are truly long-term investors and their goals are completely aligned with the old-style or 'value' form of venture capital investing" (p. 310).
19. The Chinese University of Hong Kong, for example, had total net assets of HK$8.95 billion in 2007, about HK$6.1 billion of which is investment and HK$2.7 billion is in cash and deposits. According to its 2007 Annual Report, "Taking a longer term view and in order to maximize capital appreciation, the University has formalized its investment strategy by redistributing about HK$851M into unit trusts, and invested HK$410 million in the equity of a limited partnership" (p.10). This limited partnership investment (about 7 percent of total funds) should represent an alternative investment fund.
20. See Bruton & Ahlstrom, 2003; Chau, 2007.
21. Perception Digital Ltd (HKG: 8248) was a technology venture set up by scholars who used the technology of Hong Kong University of Science and Technology. It ended its debut up 50 percent from the offer price (http://www.chinesestock.org/hklistd.asp?id=8248).
22. For an academic review of GEM, see Au et al. (2005); and for recent changes in the Growth Enterprise Market, see http://www.hkex.com.hk/rule/gemrule/gem_rupdate10_cover.htm.
23. Small Business Investment Corporations (SBICs) were private investment companies that received leveraged capital from the Small Business Administration (SBA) of the US Government. SBA set up SBICs to match the neck-breaking growth of the Soviet Union. SBICs proliferated in the 1960s, and some managers subsequently created the early generation of venture capital firms. Other countries have tried to transplant such programs to kick-start technology start-ups, such as the Enterprise Investment Scheme (EIS) in the UK and Venture Investment Support for Startups (VISS) in Singapore (Koh, 2006; Mason, 2006).
24. The rapid growth of the VC industry in recent decades has been attributed to several related events (Gompers & Lerner, 2000, pp. 8–10), one being the removal of the "prudent man" rule in 1979. It unleashed pension funds to invest in VCs; investment advisors (gatekeepers) also arose to advise and facilitate pension funds to invest.

25. See Harrison and Mason (1996) and the templates provided by the British Venture Capital Association (www.bvca.co.uk). Another useful reference is Business Angel Network (South East Asia) established by Prof. Po Kam Wong in Singapore (www.bansea.org). Recently, there has been slightly more interest in Hong Kong (Lo & Woo, 2009).
26. In the US, referring to Preston (2004, p. 6), the amended Security Act of 1933, and section 501 (c) (3) of the Internal Revenue Code. Under the Securities and Futures Ordinance in Hong Kong, "professional investors" are individuals and associates having a portfolio of not less than HK$8 million and have to be certified by an auditor or accountant.
27. http://www.straitstimes.com/Breaking%2BNews/Asia/Story/STIStory_323859.html.
28. In a recent interview, the CEO of HKSE mentioned opening a commodity futures market, recruiting companies from other countries to list, and a few others as ways to respond to future challenges. He does not refer specifically to recruiting technology firms to Hong Kong (Chan, 2008, pp. 242–243). HKSE has also stalled the study of the opening of an AIM-like "professional board" (HKEJ, 3 June 2009).

References

Amit, R., Brander, J., & Zott, C. (1998) Why do venture capital firms exist: Theory and Canadian evidence. *Journal of Business Venturing*, 13: 441–466.

Au, K., Baark, E., Chua, B. L., & Thomas, H. (2005) *Innovation Policy and High Growth Startups*. Hong Kong: CUHK Centre for Entrepreneurship.

Avnimelech, G., & Teubal, M. (2004) Targeting venture capital: Lessons from Israel's Yozma program. In A. Bartzokas, & S. Mani (eds), *Financial Systems, Corporate Investment in Innovation and Venture Capital*. Cheltenham, UK: Edward Elgar.

Bartzokas, A., & Mani, S. (2004) *Financial Systems, Corporate Investment in Innovation and Venture Capital*. Cheltenham, UK: Edward Elgar.

Berger, Suzanne and Lester, Richard K. (1997) *Made by Hong Kong*. Hong Kong: Oxford University Press.

Black, B., & Gibson, R. (1998) Venture capital and the structure of capital markets: Banks versus capital markets. *Journal of Financial Economics*, 47: 243–277.

Bruton, G. D. and Ahlstrom, D. (2003) An institutional view of China's venture capital industry: Explaining the differences between China and the West. *Journal of Business Venturing*, 18: 223.

Chan, K. C. (ed.) 陳景祥主編 (2008). *香港金融風雲35載*. 香港: 天窗出版.

Chau, W. (2007) *The Analysis of Venture Capital in Hong Kong: A Comparison of VC-Backed and Non-VC-Backed IPOs in the Growth Enterprise Market*. Australia: Doctoral Thesis, University of Newcastle.

Cheng, S. L. (2004) *Deconstructing Venture Capital Funds*. Hong Kong: Jingji Wenhua Publishing.

Child, J. (2000) Theorizing about organization cross-nationally. *Advance in International Comparative Management*, 13: 27–75.

Chinese University of Hong Kong (2007) *2007 Annual Report of Chinese University of Hong Kong*. Hong Kong: Chinese University of Hong Kong.

Gompers, P., and Lerner, J. (2004) *The Venture Capital Cycle*. Cambridge, MA: MIT Press.

Goodstadt, L. F. (2005) *Uneasy Partners: The Conflict between Public Interest and Private Profit in Hong Kong*. Hong Kong: Hong Kong University Press.

Gupta, U. (2000) *Done Deals: Venture Capitalists Tell Their Stories*. Boston: Harvard Business School Press.

Harrison, Richard, and Mason, Colin (1996) *Informal Venture Capital: Evaluating the Impact of Business Introduction Services*. London: Prentice Hall.

Jeng, L., & Wells, P. (2000) The determinants of venture capital funding: Evidence across countries. *Journal of Corporate Finance*, 6: 241–248.

Kenny, M., Han, K., & Tanaka, S. (2007) Venture capital in Asia. In H. S. Rowen, M. G. Hancock, & W. F. Miller (eds), *Making IT: The Rise of Asia in High Tech*. Stanford, CA: Stanford University Press.

Koh, W. T. H. (2006) The role of venture capital and entrepreneurship in innovation-driven economic growth. In J. E. Butler, A. Lockett, & D. Ucbasaran (eds), *Venture Capital and the Changing World of Entrepreneurship*. Greenwich, CT: Information Age Publishing.

Lo, J. Y. 盧恩澤, & Woo, P. 胡玉瑩 (2009) *Angel Investment – Successful Ways for Investors and Entrepreneurs*, Hong Kong: Hong Kong Economic Times Press.

Mason, C. M. (2006) The informal venture capital market in the United Kingdom: Adding the time dimension. In J. E. Butler, A. Lockett, & D. Ucbasaran (eds), *Venture Capital and the Changing World of Entrepreneurship*. Greenwich, CT: Information Age Publishing.

Metrick, Andrew (2007) *Venture Capital and the Finance of Innovation*. Hoboken, NJ: John Wiley & Sons, Inc.

Murmann, J. P. (2003) *Knowledge and Competitive Advantage: The Coevolution of Firms, Technology, and National Institutions*. Cambridge: Cambridge University Press.

North, D. (1990) *Institutions, Institutional Change, and Economic Performance*. New York: Cambridge University Press.

Preston, S. L. (2004) *Angel Investment Groups, Networks, and Funds: A Guidebook to Developing the Right Angel Organization for Your Community*. Kansas: Kauffman Foundation.

Sender (1993) Rippling effects. June, *Far East Economic Review*.

Wang, C. (2004) The emergence of the Singapore venture capital industry: Investment characteristics and value-added activies. In A. Bartzokas, & S. Mani (eds), *Financial Systems, Corporate Investment in Innovation and Venture Capital*. Cheltenham, UK: Edward Elgar.

White, S., Gao, J., and Zhang, W. (2005) Financing new ventures in China: System antecedents and institutionalization. *Research Policy*, 34: 894–913.

Whitley, R. D. (1992) *Business Systems in East Asia: Firms, Markets and Societies*. London: Sage.

7

Hong Kong and the Pearl River Delta: Science and Technology Cooperation

Adam Segal
Council on Foreign Relations

Science and technology, and innovation in particular, have become a critical concern of policymakers and politicians. Throughout Asia, there has been a concerted effort, through national and subnational policies, to improve technological capabilities and to create an environment of technological entrepreneurship. These policies, of course, differ by location, but there is a degree of convergence as all of the economies in the region expand financial and institutional support for small and medium-sized enterprises; foster university–industry linkages and promote university start-ups; and build cooperative agreements between state labs and private industry.

The paradox is that as innovation becomes an increasing focus of national (or subnational) policy, the process itself is becoming increasingly globalized. High-technology companies and major research universities have become untethered, searching for customers, ideas, and talent in markets around the world. High-technology companies in particular have become less "national" as they have become more global. In many industrial sectors, it is increasingly difficult to disentangle the separate national components of the value chain. The semiconductor, PC, and cell phone industries, to name just a few, involve production networks that span the Mainland, Taiwan, Hong Kong, Korea, Japan, and the United States.

How the Hong Kong Special Administrative Region (HKSAR) interacts with this larger global system of innovation, what value it adds and what benefits it can leverage, is in many ways determined by its relationship with the Chinese Mainland, and with the Pearl River Delta (PRD) in particular. Hong Kong does have links to other centers of research and development – especially the United States, Taiwan, and Japan – but getting the science and technology relationship with the PRD right is critical to Hong Kong's future.

There is already a dense web of policies, institutions, business relationships, and personal networks that tie Hong Kong and the PRD together in the areas of science, technology, and innovation, and the trend is toward even closer integration. Opportunities for joint research and development, educational exchange, business alliances, and collaborative innovation are all growing.

There also seems to be a widespread consensus that the economic context for Hong Kong is rapidly changing. For at least the last 20 years, the relationship could be summarized as "front shop, back factory" (*qian dian hou chang*), or the location of production in the PRD and important services, such as financial services, marketing, design, insurance, communications, and logistics, in HKSAR. But the PRD, while still China's manufacturing center, faces increased competition from other places in the Mainland, including the Yangtze River region and the Beijing–Tianjin corridor, as well as other regions in Asia such as the Singapore–Malaysia–Indonesia triangle. Most importantly, the PRD is no longer content to be the back factory. The region has committed to moving up the value chain, spending more on R&D, upgrading manufacturing capabilities, and creating clusters of universities, R&D institutes, and start-up firms in such places as the Science Park in Shenzhen, Songshan Lake in Dongguan, and Bio-Island in Guangzhou.

While there is a shared diagnosis of the situation, it is difficult to escape the impression that there is no shared, coherent vision of what the ultimate objective is – of what Hong Kong should be doing about the changing economic environment. Given the widespread view that Hong Kong's historical success has been based on a market-driven, laissez-faire approach to economic development, it is not surprising that science policy does not have a great deal of institutional or public support. Some argue that the government's inability, or unwillingness, to pursue a more coherent innovation policy reflects the public's general lack of interest; others reverse the causality, seeing government's failure to argue forcefully for the need for innovation policy as being, in part, behind public apathy. These critics worry that Hong Kong's traditional "hands-off" approach is fundamentally incompatible with what needs to be done to build a knowledge economy in the HKSAR.

Some of this uncertainty about how to respond to new challenges emerges from larger political or ideological debates about Hong Kong's future – how close should the relationship be between Hong Kong and the PRD, and how hard should Hong Kong work to maintain some sense of separateness from the Mainland? Should the stress be on "One Country" or on "Two Systems"? There are those who fear that Hong Kong is not integrating with the Mainland fast enough, and will soon be marginalized by both the rapid growth in China and the loss of Hong Kong's role as the only door into China. The ambitious development strategy outlined in Guangdong's 11th Five Year Plan, for example, provoked a great deal of soul-searching in the HKSAR, and former Chief Secretary for Administration Rafael Hui Si-yan created a small storm when he warned that Hong Kong could be left on the sidelines.[1]

In December 2008, the National Development and Reform Commission (NDRC) announced an even more ambitious plan for the PRD, *The Outline of the Plan for the Reform and Development of the Pearl River Delta (2008–2020)*.[2] The plan not only provides HK$5.68 billion for the construction of a bridge linking Hong Kong, Macau and Zhuhai, but also sets the goal of shifting the Delta from a low-cost manufacturing base to a locale for high-end manufacturing, modern services, and finance. The region will develop a:

> strategic orientation toward high-end development, build a new stronghold for independent innovation, forge a number of advanced manufacturing bases that rank high among their world counterparts in both scale and quality, foster a batch of internationally competitive world-class enterprises and brands, develop a system of service industries to match Hong Kong as an international financial center, and develop into an international center for shipping, logistics, trade, conferences and exhibition, tourism, and innovation that has a different positioning from Hong Kong and Macao.

Manufacturing powered by high technology, according to the plan, will generate at least 30 percent of the region's total industrial output by 2020. In the end, Guangdong "will pursue convergence with Hong Kong and Macau in terms of urban planning, rail transit networks, information networks, energy base networks and urban water supply."

While many will be pleased with this new impetus to integration, for some the prospect of this convergence of the HKSAR and the PRD is an issue. The journalist Philip Bowring argues that it is Hong Kong's uniqueness that in fact makes it competitive, and so integration should proceed slowly; Hong Kong "has a struggle on its hands to retain an identity which allows it to follow policies to help sustain its position as one of the world's richest cities, which necessarily means revisiting pressures to force the pace of integration with a still much poorer Mainland."[3]

In many ways, the HKSAR is archetypal of the interactions between globalization and regionalization – the SAR embodies, and benefits from, both the distribution of production, technology, and capital across global borders as well as the forces of interdependence that tightly tie economic regions together as units of competition. With the financial crisis of 2008–2010, it is highly possible that the world will move in a more regional direction, not an even more global one – Europe, North America, and Asia will turn inward and raise trade barriers against each other. If this is the future, then it is likely that the HKSAR will become even more dependent on the PRD and the People's Republic of China.

There is an opposing argument to be made, however. There can be real economic and political benefits to be leveraged from Hong Kong keeping close economic ties with the rest of the world. The question, for the HKSAR and China, is what degree of autonomy is useful to both sides?

In addition to these larger questions, there are more structural problems of coordination between two different political systems and multiple actors, and of finding complementarities among several economies all at different levels of development and with different comparative advantages. There are many actors involved, spread across several legal jurisdictions, and all are not pursuing the same interests at the same time. Even when actors agree on common goals, implementation remains difficult. Perhaps more worrying for the people of Hong Kong, some of the reporting on the 12-year plan for the PRD suggests that the HKSAR had a limited role in the planning process. "Only if the government proactively participates in relevant plans, can Hong Kong make full use of its advantages and achieve a win-win effect with the Mainland," suggested an editorial in *Ming Pao*. "If it plays no role in the planning process and does not participate actively, the outcome will be 'being planned [by the Mainland].' "[4]

There are specific policies that can be adopted to address the issue of coordination. But before these policies are tackled, the larger political issues need to be addressed – issues that need to be discussed beyond the narrow range of innovation policy. One of the biggest barriers to improved science and technology (S&T) coordination, for example, is, as Eric Thun also notes in his chapter, that policymakers continue to try and distinguish between Hong Kong and PRD companies. Yet it no longer makes much sense to distinguish firms based on ownership – Mainland firms can list on the Hong Kong stock market, Hong Kong firms no longer need to have a front office in the HKSAR. In addition, Hong Kong's positioning as a service center or as a platform for intellectual property depends in large part on the quality of the manufacturing in the PRD. Or, as a Dongguan official put it: "We tell our counterparts in Hong Kong: You are not helping Dongguan, you are helping your own companies, and you are helping yourself."[5]

There are some seemingly simply policy solutions to this problem. But for Hong Kong to really move forward as an innovation hub, it may be necessary to have greater integration with the PRD. Exploiting all the opportunities offered by the Mainland may require the HKSAR to sacrifice a degree of its autonomy. In effect, the globalization–regionalization conundrum must be addressed head on. But this is as much a political decision as it is a policy one.

1. The current state of HKSAR and PRD S&T relations

There already exists a dense web of policies, consultative mechanisms, institutions, and personal networks that tie Hong Kong to the PRD in the areas of science, technology, and innovation. Since January 2006 and the third phase of the Mainland and Hong Kong Closer Economic Partnership Arrangement (CEPA), all Hong Kong goods can be exported duty-free to China. CEPA also includes preferential access to the Mainland for Hong Kong services companies, including patent and trademark agencies and those offering information technology services. Still, many argue that CEPA's most important contribution has not been the tariff reduction, but instead the relaxation of travel restrictions on Mainland travelers visiting Hong Kong. The number of Mainland Chinese visitors jumped from 3.79 million in 2000 to 13.6 million in 2006, spurring a recovery in the retail and tourism sectors.[6]

Two recent policy initiatives address the question of collaborative R&D and innovation directly. The Guangdong–Hong Kong Technology Cooperation Funding Scheme (TCFS) supports collaborative research. All proposals must demonstrate an element of Guangdong–Hong Kong cooperation. Universities, research institutes, trade and industry associations, professional bodies, and local companies can all apply for funding under three research schemes: *Platform*, which includes publicly funded R&D institutes and trade associations; *Collaborating*, which is private companies with local publicly funded research institutes; and *Company*, which is private companies without publicly funded research institutes.[7]

The applications are solicited and vetted either by one of the five research centers established in 2006 (automobile, nanotechnology and new material, information technology and communications, logistics, and textiles), or by the Innovation and Technology Commission (ITC). As of July 2008, 26 research and development projects have been approved, with total funding amounting to around HK$120 million.

In February 2009, as part of the TCFS, HKSAR and Shenzhen announced they would jointly fund eight applied research and development projects. The Hong Kong government will contribute HK$36.44 million to R&D projects in electric vehicles, battery technologies, radio frequency identification (RFID) applications, and energy and green technologies.[8]

The Shenzhen/Hong Kong Innovation Circle is a broad framework for increasing collaboration. Initiative for the plan, at least according to one report, came from the Shenzhen government. A 2006 Shenzhen government document called innovation a strategic goal, and subsequently Shenzhen officials pressed the Hong Kong Trade Development Council, Hong Kong Science and Technology Park, Chinese University of Hong Kong, and the Hong Kong University of Science and Technology for support. The first major project under the Innovation Circle umbrella, announced in May 2008, is a joint project of Shenzhen, Hong Kong Science and Technology Park, and DuPont to establish a Solar Energy R&D Support Center.[9]

The Innovation and Technology Commission is involved in several cooperation mechanisms including: the Mainland/Hong Kong Science and Technology Cooperation Committee, the Pan-PRD Joint Conference on Regional Cooperation in Science and Technology, the Guangdong/Hong Kong Expert Group on Cooperation in Innovation and Technology, and the Steering Group on Shenzhen/Hong Kong Cooperation in Innovation and Technology. The Hong Kong/Guangdong Expert Group on Cooperation in Informatization serves a similar consultative role, though the topics tend to be of a more applied nature and include the use of RFID in logistics and radio spectrum management, as well as joint development in interoperability of open source software, next generation networks, and wireless and mobile technology.

1.1. Institutions

Dotting the landscape are a number of institutions that link the two regions together. The PKU-HKUST Shenzhen-Hong Kong Institution was established in August 1999 as a joint venture of the Shenzhen Municipal Government, Peking University (PKU) and the Hong Kong University of

Science and Technology (HKUST). It is located in the Shenzhen Hi-tech Industrial Park and acts as incubator for high-tech professionals in the Shenzhen–Hong Kong bay area. Through its 2005–2020 Strategic Plan, HKUST also plans to develop R&D and spin-off activities in five areas – nanotechnology, biotechnology, electronics, wireless, and sustainable development – in Shenzhen.

The Nansha IT Park is a project of Hong Kong University of Science and Technology, Fok Ying Tung Foundation, and the Guangzhou government, designed to foster high-tech industries in the region. In September 2008, HKUST opened a new graduate school in the Park. In addition, located in the Shenzhen High-tech Industrial Park, the Shenzhen-Hong Kong Productivity Foundation supports Hong Kong industries operating in Shenzhen.

The Hong Kong Productivity Council supports Hong Kong enterprises operating in the PRD through three subsidiaries in Guangzhou, Shenzhen, and Dongguan. These three subsidiaries provide training, business consultancy, and IT industry support services. Opened in 2007, the Shenzhen-Hong Kong Productivity Foundation focuses on product innovation, especially in the areas of environmental technology, electronics and automotive technology, and software and digital entertainment.

1.2. People

Parallel to these institutional frameworks, frequent personal contacts and meetings occur throughout the region. Members of science bureaus in Dongguan, Zhuhai, and Shenzhen all spoke of numerous contacts with members of the ITC. Conversely, the ITC reports collaborative arrangements with the Mainland at different levels, including the Ministry of Science and Technology, the Ministry of Information Industry, Guangdong Provincial Department of Science and Technology, the Shenzhen Municipal Government, and various provinces in the pan-PRD region.

The Nansha Science and Technology Forum brings policymakers, scientific experts, and business leaders together to discuss the region's technological, social, and economic development. Professional associations throughout the region, including the Internet Professional Association and the Chinese Software Professionals Association, hold networking meetings, conferences on special topics, and business plan competitions.

Joint academic research and publications are also critical. HKUST aids Shenzhen enterprises seeking to upgrade manufacturing capabilities, has established a joint R&D center with Huawei, and undertakes projects under the National Basic Research Program (also known as the 973 Program), organized by the Chinese Ministry of Science and Technology. For scholars on the Mainland, Hong Kong academics ranked third in 2000 as co-authors on research papers (339 joint papers with Chinese scientists as first author), after the United States (587) and Japan (566).[10]

Most importantly, the HKSAR is connected to the PRD through inter- and intra-firm connections. The Mainland is a major source for, and destination of, Hong Kong high-technology products. In 2007, 61 percent of all IT equipment exports were to the Mainland; slightly more than 70 percent of IT equipment re-exports from Hong Kong were originally from the Mainland. For telecom equipment, 38 percent of exports were to the Mainland ; re-exports from the Mainland make up close to 80 percent of re-exports from Hong Kong.[11]

These connections tend to fall under the traditional "front shop, back factory" division of labor. Component design and application, quality control, project management services and logistics support, software development and systems integration, and marketing and licensing all occur in the HKSAR. Labor-intensive manufacturing processes are carried out in the PRD, as are, increasingly, some research and development. Vtech, for example, moved manufacturing to Dongguan, and R&D to Shenzhen, after the company decided it was too difficult to find engineers in HKSAR: "fewer and fewer people in Hong Kong can do this. And engineers in Shenzhen are about half the cost."[12] Product management, the development of a roadmap for the company, and interface with customers remained in Hong Kong, and Vtech expected this division of labor to stay this way.

2. Tightening the link

The central organizing principle for future cooperation, for many on both sides of the border, is that Hong Kong will act as R&D and service center as well as IP platform for the Mainland. As Mainland companies move up the value chain, they will need help acquiring market intelligence and the rights to use technology products from abroad, as well as developing branding and marketing strategies, and it is in these spaces where Hong Kong will play a role. Most studies seem to support at least the structural basis for this relationship, as the PRD is said to lack service companies.[13] S&T officials in Dongguan said, "When we think about cooperation with HKSAR we think of services as well as the city's orientation toward the West."[14]

Yet despite the widespread acceptance of this division, there are some doubts about how effectively the cooperation is developing. The same Dongguan officials who told me of Hong Kong's importance as a service center, also stressed that this role was not technology oriented, but in the realm of finance and marketing. Some question whether Hong Kong has or will ever have the R&D capabilities needed to play an active role in the PRD's upgrading. Several of the studies in this volume, especially those by Mowery and Wong, suggest that HKSAR universities are not keeping pace with the increasing applied R&D capabilities of Mainland universities. Yeh and Xu come to a particularly pessimistic conclusion: "Because of its own low technological level, Hong Kong's role in supporting the PRD's high tech development is minimal, if not totally impossible."[15]

At the same time, there is anecdotal evidence that building Hong Kong as an intellectual property platform may not be easy, as multinational companies could find their own ways to work around the problem of the lack of IP protection on the Mainland, cutting the HKSAR out of the equation. As a manager at STMicroelectronics told Douglas Fuller, the company, after setting up a design center in Shenzhen, went to Shanghai rather than Hong Kong. This suggests that while IP protection is better in Hong Kong and the Hong Kong Science and Technology Park (HKSTP) offers better services for integrated circuit (IC) design, it may not matter much for firms that have operated successfully in the Mainland. They may have developed their own best practices – dividing processes into component parts, keeping the most valuable IP at home – which makes Hong Kong less attractive.

Despite the wide range of interactions occurring between the PRD and the HKSAR, much of the cooperation is seen as not fully utilized, to be in its early stages, or not effective. Or, in the words of one Shenzhen S&T official: "The scale is not large enough. Cooperation is broad but not deep."[16] The challenges fall into three categories: leadership, coordination, and competition.

2.1. Leadership

There is a perception in China that no one in the HKSAR government "owns" the issue of technological cooperation and so there is no one on the Hong Kong side to push things through when good ideas are brought up. As one S&T official in Dongguan put it: "The ITC is interested in many of the things we suggest but the problem is no one listens to them. They have to get leaders above interested."

Some on the Mainland suggested that this lack of leadership comes from the desire to protect "One Country, Two Systems," making the HKSAR very slow and very conservative. As noted above, there is, at least in some corners, a sense that Hong Kong played a very passive role in the NDRC's plan for the reform and development of the PRD. In the realm of science and technology, the default response to any new policy initiative, according to one Dongguan official, is "We cannot do this."

The lack of leadership on the Hong Kong side means no one is prepared to address the most pressing issue: the mismatch between what PRD companies want (or at least what PRD officials say companies want) and what the HKSAR can or will provide in the way of investment. The prohibition on Hong Kong money being spent outside the 1000 sq km area of HKSAR, or, in the phrase that came up many times in interviews, "money can't cross the river," may be the biggest barrier to deeper cooperation. The issue here, as mentioned in Eric Thun's chapter, is that policy is now focused on ownership – nominally Hong Kong or Mainland firms – and not on improving the competitiveness of the HKSAR and PRD together.

2.2. Coordination

This problem of leadership is compounded by the everyday problems of working across two systems and dealing with different educational systems, as well as tax, labor, and visa laws. The planning timeframe is different, with the HKSAR producing new budgets and policies every year and the PRD engaging in 5- and 12-year plans, although Hong Kong has moved to conducting some longer-term planning with its "Hong Kong's 2030 Vision and Strategy." In addition, not surprisingly, neither the Hong Kong nor the PRD side speaks with one voice; they each have their own coordination problems.

There is a redundancy in the plans for science and technology in each of the PRD provinces. Each of the provinces pursues its own interest, with a great deal of overlap in large-scale projects and basic research. According to Lu Jianbao, "each provincial policy possesses strong regional protectionism, which obstructs the exchange among factors of technology innovation, the usage of resources to a great degree, and does not benefit the smooth operation of the region's innovation system."[17]

For much of the first decade of the twenty-first century, it did not appear that Beijing was willing or able to resolve these coordination problems. At least until December 2008, there was no evidence that Hong Kong and its innovative capabilities weighed heavily in the thinking of decision-makers, policy analysts, or entrepreneurs in Beijing. It was only in 2007 that the Ministry of Science and Technology (MOST) agreed to accept applications from Hong Kong universities and research institutions to set up State Key Laboratories in Hong Kong. The program, which began on the Mainland in 1984, engages local universities and research institutions in the key national technology programs. ITC, along with the Research Grants Council, is now reviewing 17 applications, with results to be announced in 2008.

The NDRC, in addition, has many programs to upgrade SMEs, including active programs in the PRD, but none of these include activities for Hong Kong or cooperation with Hong Kong. According to an interview conducted by Douglas Fuller, NDRC officials did not seem to think this line of cooperation would be explored much in the future either.

The NDRC's plan may signal a change. The guidelines explicitly mention the creation of a "new regional layout of innovation along the Guangzhou-Shenzhen-Hong Kong axis" through the deepening of the science and technology cooperation among Guangdong, Hong Kong, and Macao, establishing zones for joint innovation and strengthening. It also states that Beijing will encourage Guangdong, Macao, and Hong Kong to increase their consultation and to formulate regional cooperation plans.

2.3. Competition

Finally, there is a clear element of economic competition between the HKSAR and the PRD. Today, many of the cities in the PRD are thinking of moving into areas that are, or could be, Hong Kong's competitive advantage. The NDRC's 12-year plan speaks of the PRD developing modern service industries, logistics, information services, and science and technology services, with "in-depth cooperation with Hong Kong and Macao."

The question is whether their ambitions will come at the expense of Hong Kong. Several of the chapters in this book suggest exploiting better one of Hong Kong's clearest competitive advantages, especially vis-à-vis the Mainland : its excellent universities. The NDRC's plan sets the goal of Guangzhou, Shenzhen, and Zhuhai establishing "cooperative higher education institutions with three to five famous foreign universities and the region will develop one to two universities that are first class in China and enjoy a leading status in the world."

Writing about Shenzhen, Bowring argues that momentum lies to the north: "Shenzhen has set its sights on becoming an 'international city.' This is supposed to be achieved by strengthening so-called 'cooperation' with Hong Kong in the very areas where Hong Kong now excels, finance and logistics. Given the political dynamics, Shenzhen would be more likely to absorb Hong Kong than vice versa."[18]

Shenzhen is also moving aggressively to develop its indigenous capabilities and develop S&T products. There is a clear desire to change the industrial nature of Shenzhen, to move from "processed in Shenzhen" to "made in" and eventually "designed in" Shenzhen. In the 11th Five Year Plan, spending on R&D will total RMB 100 billion, with the government share at 10 percent. In 2006, Shenzhen spent 3.4 percent of its gross domestic product on research programs – the highest percentage in China – and the city trails only Beijing and Jiangsu in absolute spending. "Our spending on technology research is now second only to our education spending. We are very serious about it," said Zhang Lailin, deputy director of the Shenzhen Trade Development Bureau.[19] One local official put it even more bluntly: "Shenzhen has to invest in R&D or it will die."[20]

During interviews in Shenzhen, it was easy to note an explicit critique of Hong Kong for not being as aggressive as Taiwan and Singapore about pursuing technological innovation through industrial policy, as well as a somewhat lecturing tone regarding what Shenzhen had accomplished while the HKSAR had stood still. In fact, several interviewees explained that people in Hong Kong did not really understand what was happening in their backyard, either because they still looked down on the Mainland, or because they were "too close." Hong Kong businessmen make the trip in a day and so, according to one Dongguan official, "are not integrated into local business community. The Taiwanese are much better at this."

3. The way forward

Given these difficulties in leadership, coordination, and cooperation, the HKSAR could move forward in five areas: platforms, resources, borders, institutions and leadership, and diversification. As noted at the beginning of this chapter, however, the success of many of these actions will be determined by a larger, inherently more political decision to treat the PRD and HKSAR as one economic entity – to view new technological capabilities located in the PRD as critical to Hong Kong's competitiveness.

3.1. Common platforms

Policies focused on supporting specific technologies are unlikely to be as successful as those focused on the development of common platforms. These policies avoid many of the typical shortcomings of "pick the winner" industrial policy, and play to the natural division of labor between HKSAR and PRD. There has been some movement toward the creation of common accreditation and educational standards. United International College, situated in Zhuhai and jointly founded by Beijing Normal University and Hong Kong Baptist University, is the first full-scale cooperation in higher education between the Mainland and Hong Kong. The goal is to develop a new model of liberal arts education for the Mainland, but more needs to be done systemically to bring Hong Kong and Mainland university systems into greater sync, at the same time as Hong Kong universities move to 4-year degrees by 2012.

The most obvious area of cooperation is the development of new technology standards. Hong Kong Applied Science & Technology Research Institute (ASTRI) is already involved in a joint project on developing standards for home networking and applications as well as high-speed wireless, and there are various discussions by the business and professional associations on RFID and other technologies. The Guangdong–Hong Kong RFID Industry Cooperation Memorandum was signed in June 2007.

The challenge for Hong Kong, both at the association and individual firm level, is how to participate in the development of new Chinese standards without isolating itself from international standards and global markets. In some instances, and in the case of WAPI (WLAN Authentication and Privacy Infrastructure, a competitor to WiFi) in particular, Chinese standards have been exclusionary, developed without international participation, and viewed by many multinationals as a tool to force technology transfer.

In addition, Hong Kong already operates as an intellectual property rights (IPR) platform – it attracts multinational corporations (MNCs) that want to operate in the Mainland but fear the high level of IPR theft. It can further leverage this competitive advantage, especially with a concerted effort to reach out to smaller firms that want to do business in China but that do not have the confidence to enter the market on their own, providing a one-stop clearing house of best practices, trusted partners, and due diligence. In addition to hosting a website that provides advice to companies wanting to export to China (www.export.gov/china), the US Department of Commerce, in cooperation with the American Bar Association, the National Association of Manufacturers, and the American Chamber of Commerce in China, runs the China Intellectual Property Rights Advisory Program, which offers free consultation on IPR issues to American small and medium-sized enterprises. Hong Kong could offer a similar service.

As with standards policy, there are possible pitfalls ahead. There could be a tension between protecting Hong Kong's reputation as a location with a strong IPR regime and its increasing collaboration with Mainland firms, who may not be the most protective of foreign IPR. Hong Kong officials, as well as its business association, will have to monitor closely the state of Hong Kong's international reputation.

Finally, there should be a central directory of all the S&T cooperation occurring between the two regions. In 2007, the Hong Kong and Shenzhen governments set up a website to promote the sharing of equipment, labs, and professors at technological institutes. This could be expanded to include the entire region and would serve as a clearing house for all projects, which might help identify potential areas of future cooperation and prevent redundancy.

3.2. Resources

Regulations preventing "money from crossing river" should be reconsidered in light of the link between the technological capability of the HKSAR and the manufacturing quality of firms operating in the PRD, no matter what the ownership. Currently, the government is considering relaxing the geographical restrictions on the University–Industry Collaboration Program, which supports commercial R&D projects undertaken by private companies in collaboration with local universities. In addition, a PRD component could be added to other ITC funding programs – the Small Entrepreneur Research Assistance Program could include Mainland partners, for example.

Hong Kong officials could also be more assertive about pushing for the inclusion of local universities and research institutes into Mainland funding programs. As mentioned earlier, it was only very recently that money from the 973 Program began going to Hong Kong institutes, and, as Joseph Wong's chapter notes, although biotech labs at HKU have received "key state lab" recognition from the Mainland, they have yet to receive any money from the PRC government. Similarly, Douglas Fuller notes that Hong Kong received essentially no support or funding from the central government for IC development plans.

3.3. Borders

There remain some geographic barriers to cooperation, with numerous interviews mentioning the need to improve the flow of people and visa processes for people from PRD coming into HKSAR, and vice versa. The visa process for Mainland engineers and scientists needs to be rationalized. Also, as mentioned by David Mowery, there is a need to open Hong Kong universities to greater undergraduate enrollment from the Mainland.

3.4. Institutions and leadership

Many of the issues of coordination – as well as addressing the perception that the HKSAR is essentially reacting to the development plans of Shenzhen – might be resolved by a more vocal and public trumpeting of the goal of S&T collaboration. Hong Kong may be required to strengthen the role of ITC to fulfill this goal of coordination, or possibly to shift the responsibility for HKSAR–PRD S&T cooperation to the secretary in order to give it a higher profile and greater bureaucratic power. At the very least, there need to be clear and consistent statements at the top – from the chief executive and others – regarding the importance of innovation to Hong Kong's future, and of collaborative innovation in particular.

At the very least, the HKSAR needs to seize the initiative on the issue. This means not allowing the "Shenzhen/Hong Kong Innovation Circle" to be defined primarily by talking and plan making, a fate that seems to have befallen many of the other collaborative projects that preceded it. The Solar Energy R&D Support Center is a good start, and it needs to be followed with additional projects.

3.5. Diversification

S&T connections with the PRD are clearly the most important, but Hong Kong also has to strengthen ties with other centers of innovation. In fact, the more tightly linked to the rest of the world Hong Kong is, the more valuable it is to the PRD; the more closely tied to the PRD, the more attractive Hong Kong is to the rest of the world. The local universities, HKSTP, and the research institutes need to strengthen cooperation with Taiwan, Southeast Asia, Japan, Korea, and India. The government should also hold a HKSAR–ASEAN Technology Summit (HKSAR–India, etc.), designed to provide an opportunity for R&D institutions, academia, industries, and governments of ASEAN (Association of South-East Asian Nations) countries and HKSAR to interface.

Hong Kong will also want to leverage more informal networks and try to attract young technology entrepreneurs from India, Southeast Asia, Europe, and the United States through subsidized space at Cyberport and HKSTP or other benefits, and make it easier for scientists and engineer from these countries to obtain visas. As Douglas Fuller notes (see this volume, Chapter 10), few Hong Kong engineers in Silicon Valley have been lured back because of the lack of an existing viable tech sector, lack of venture capital, and limited government support which gives few incentives to return. The HKSAR may also consider assisting in the development of an equivalent of The Indus Entrepreneurs (TiE) and other expatriate groups for Hong Kong entrepreneurs located around the world.

4. Conclusion

The task ahead for Hong Kong is not an easy one. Coordination between the HKSAR and the PRD is bound to be complicated given the multitude of actors and interests involved. Moreover, especially in the shadow of the NDRC 12-year plan, Hong Kong will have to work hard to fight the impression that it is no longer in the driver's seat, that the initiative and energy for regional development lie in the PRD.

The difficulty for Hong Kong is more than a coordination problem. Instead, it is the larger question of how much autonomy is good for Hong Kong and for the Mainland. There seems little doubt that Hong Kong's economic future is tied to the PRD. If Hong Kong truly wants to develop its own innovative capabilities, then it will have to rely on the resources and talent of the PRD. Closer collaboration will be critical. Still, a relatively autonomous Hong Kong – one that acts as a filter between the PRC and the rest of the world, that retains a strong independent financial capability, and builds on its position as a regional center of educational excellence – could be a major source of strength for China. Creating that autonomy, however, is not so much a policy process as a political one.

Notes

1. Lau Nai-keung (2006) "Hong Kong Must Adjust Development Strategy," *China Daily*, 12 April.
2. Available at: http://www.newsgd.com/government/govpress/content/2009-02/27/content_4947625.htm.

3. Philip Bowring (2008) "At Risk: Hong Kong's Identity – and Prosperity," *Hong Kong Journal*, Winter. Available at: http://www.hkjournal.org/archive/2008_winter/3.htm.
4. See, for example, "Playing no role in the Planning process can only bring 'being planned': Hong Kong government needs to actively participated in planning that involves Hong Kong," *Ming Pao*, Jan 2, 2009.
5. Author interview, Dongguan, 10 June 2008.
6. Research Institute of Economy, Trade and Industry (2007) "Tenth Anniversary of Hong Kong's Return to China – Riding out the tough times and readying for a new leap," 29 June. Available at: http://www.rieti.go.jp/en/china/07062901.html.
7. *Guide to the Innovation and Technology Fund, Volume I: Innovation and Technology Support Programme* (May 2009). Available at: http://www.itf.gov.hk/eng/Forms/itf-itsp-guide_2009.pdf.
8. George Leopold (2009) "China Announces R&D Projects," *EE Times*, 6 February.
9. "Shenzhen, HK to collaborate in R&D," November 2009. Available at: http://www.news.gov.hk/en/category/businessandfinance/091116/html/091116en03004.htm.
10. Wu Yishan et al. (2004) "China Scientific and Technical Papers and Citations (CSTPC): History, Impact and Outlook," *Scientometrics*, Vol. 60, No. 3, 385, 397.
11. TDC Research, Shenzhen-Hong Kong Technology Industry Cooperation 2007.
12. Author interview, Vtech, 11 June 2008.
13. See, for example, Jianbao Chen (2006) "Research on Pearl River Delta Economy's Technological Innovation System Construction," *Pearl River Economy*, June.
14. Author interview, Dongguan, 10 June 2008.
15. Anthony Yeh and Jiang Xu (2006) "Turning of the Dragon Head: Changing Role of Hong Kong in the Regional Development of the PRD," in Anthony G. O. Yeh (ed.), *Developing a Competitive PRD in South China Under One Country–Two Systems* (Hong Kong: Hong Kong University Press), 63–96.
16. Author interview, Shenzhen, 9 June 2008.
17. Jianbao Chen (2006) "Research on Pearl River Delta Economy's Technological Innovation System Construction," *Pearl River Economy*, June.
18. Phillip Bowring (2008) "Two Threats to Hong Kong," *International Herald Tribune*, 16 July.
19. Quoted in Chow Chung-yan (2006) "$2HK billion boost for delta technology," *South China Morning Post*, 14 October.
20. Author interview, Shenzhen, 9 June 2008.

8
Manufacturing for a Post-Manufacturing City

Eric Thun
Oxford University

There was a time when Hong Kong was associated with cheap manufactured goods such as toys, low-end electronics, and garments, but this was well over two decades ago. In the last two decades, Hong Kong has undergone a remarkable transformation. The factories moved north when rising costs in Hong Kong made manufacturing uncompetitive, and the territory became a high-end service center dominated by gleaming office towers and shopping malls. Although the transformation of Hong Kong has been remarkably successful, the ever increasing capabilities within Mainland China create the potential that the competitive advantage of Hong Kong will decline over time.

The premise of this book is that Hong Kong should be wary of relying too heavily on its core strengths, and that there is an urgent need to cultivate a "core" of technology competencies. In this context, the subject of general manufacturing and the research and development (R&D) that supports it, might seem like a strange topic for inclusion. General manufacturing is a catch-all category that is defined primarily by *not* being high-tech. It does not include any of the activities that are normally associated with a center of innovation; it is not biotech, IT, software, integrated circuits, or any of the other sectors that are likely to quicken the pulse of a techno-nationalist. It is mid-technology industries (or activities within an industry), those that involve innovation capabilities that are not based on pure science, but also involve production and project execution skills that require more than low-cost labor. Process and product innovation in these industries tends to be incremental rather than radical, and technology is widely available (for a price) from global suppliers (Berger and Lester, 1997, 325). These are the "traditional" manufacturing industries that Hong Kong has been abandoning for the past two decades. Given that the firms involved in these activities are located overwhelmingly in Mainland China, and the number of people they employ in Hong Kong is kept to a bare minimum, supporting these industries would appear to have very little impact on Hong Kong itself.

Although supporting general manufacturing might appear to be anachronistic in a post-manufacturing city, in this chapter I will argue that, much to the contrary, support for traditional manufacturing should be a crucial element of Hong Kong's upgrading strategy. The argument will be presented in three parts. The first section of the chapter explains the linkages between Hong Kong and the manufacturing that takes place in the Pearl River Delta (PRD). This has been the subject of extensive research, particularly in the mid-1990s, and, like some of these studies, I will also argue that manufacturing in the PRD is crucial to the economic health of Hong Kong because it supports producer-related service in Hong Kong. There have been two key changes in the last decade, however: (1) the continued blurring of the distinction between a Hong Kong firm and a Mainland firm, and (2) the urgency of the upgrading challenge in the PRD as costs rise in the region. As a result of these two trends, I will argue that it makes little sense to distinguish between Hong Kong and Mainland firms in the PRD; the objective should be to promote the economic well-being of the regional economy (of which Hong Kong is the core service center). The second section of the paper analyzes potential strategies for upgrading in manufacturing and the role of government-funded research centers in this process. The final section discusses the policy implications.

1. Hong Kong and the regional economy

Hong Kong is one of the most service-oriented economies in the world, with the tertiary sector accounting for 90.7 percent of GDP in 2005 (Enright and Scott, 2007, 66). This was not always the case. Hong Kong made a rapid transition from an entrepôt economy to a manufacturing economy after the United Nations imposed an embargo on trade with China in June 1951, and within a decade the manufacturing sector was employing 40 percent of the workforce and contributing one-fourth of GDP (Berger and Lester, 1997, 19). Industrialization in Hong Kong was driven by small and medium-sized enterprises (SMEs) that began with labor-intensive products that were exported to foreign markets and gradually upgraded to more sophisticated products and markets; the electronics industry barely existed in 1960, but produced almost one-fifth of Hong Kong's industrial output in 1980. Three decades after World War II, Hong Kong was the largest exporter of manufactured goods in the developing world (Berger and Lester, 1997, 20–21). But the success of manufacturing in Hong Kong sowed the seeds of its own demise, and as wages and the cost of land increased, manufacturing operations shifted to the PRD. By the beginning of the twenty-first century manufacturing in Hong Kong was producing only 5.9 percent of GDP and employing 10.4 percent of the workforce (Hong Kong Policy Research Institute, 2003).

Despite the decline of manufacturing within Hong Kong, numerous studies in the 1990s made the point that manufacturing continued to be of vital importance to the Hong Kong economy. *Made by Hong Kong,* a study conducted by a team of Massachusetts Institute of Technology (MIT) researchers, argued that distinguishing between manufacturing and service activities was a statistical fiction because large numbers of service firms in Hong Kong were actually engaged in manufacturing in China and many other service firms were heavily reliant on the manufacturing firms that relied upon their services. The important distinction was between consumer services (such as tourism, restaurants, retail trade, retail banking, health care, etc.) and producer services that provide intermediate inputs in production value chains, such as design, logistics, and finance (Berger and Lester, 1997, 28–29). Tao and Wong, making a similar distinction, attempted to divide out the different types of services, and document a gradual increase in producer-related services as a percentage of GDP over the course of the 1990s (2002, 2349). By 2000, manufacturing in Hong Kong had declined to only 5 percent of real GDP, but as Table 8.1 indicates, producer-related services had increased to almost 50 percent of real GDP. In short, the steady decline of manufacturing in Hong Kong during the 1980s, 1990s was mirrored by the increasing importance of manufacturing by Hong Kong firms in the PRD and service activities within Hong Kong to support these operations.

If the manufacturing operations of Hong Kong firms in the PRD support higher value-added service activities in Hong Kong, the primary public policy challenge is ensuring that Hong Kong retains its competitive advantage in both services (in Hong Kong) and manufacturing (in China). The former requires a strengthening of the infrastructure (both physical and institutional) that supports producer services. The latter requires the development of capabilities in industry, government, and the education community that will allow firms to move from low-end production into higher value-added activities. *Made by Hong Kong,*

Table 8.1 Producer services in Hong Kong

	1980	1985	1990	1995	2000	2005
			Share % of Real GDP			
Manufacturing	14.6%	13.5%	14.0%	8.7%	5.4%	3.3%
Total Services	79.3%	79.5%	78.5%	83.6%	86.6%	90.8%
Producer Services	36.1%	31.6%	34.7%	41.7%	45.8%	52.8%
Consumer Services	34.9%	38.4%	35.9%	33.6%	33.3%	31.4%
Government Services	8.3%	9.6%	7.9%	8.3%	7.6%	6.6%
Others	6.1%	6.9%	7.5%	7.8%	8.0%	5.9%

Source: CEIC, YCR Wong, Z Tao and CS Chan, *An Economic Study of Hong Kong's Producer Service Sector and Its Role in Supporting Manufacturing*, funded by Industrial Support Fund, May 2000, 112 pages; and Z Tao and YCR Wong, "Hong Kong: From an Industrialized City to a Center of Manufacturing-Related Services," *Urban Studies*, Vol. 39, No. 12, 2002, pp. 2345–2358. HKCER, 2007, 21.

for instance, provided detailed recommendations on the changes and investments that firms must make to create the capability to move into design activity or brand-name production, that educational institutions must make to ensure that they are providing the supply of appropriately trained managers, designers, and engineers that firms require, and that government must make to ensure that officials have the technical capabilities required to oversee this industrial transformation (1997, chapter 7).

How has the situation changed over the course of the last decade? First, it is becoming increasingly difficult to distinguish a Hong Kong operation in the PRD from a domestic Mainland operation. As has been the case for a long time, many firms are traders in Hong Kong and manufacturers in the Mainland – there were 123,000 manufacturing and trading companies in Hong Kong in 2002 and 63,000 (or 53 percent) of these firms had manufacturing operations on the Mainland (HKCER, 2007, 15) – but there have been subtle shifts in the nature of these linkages. A survey of firms located in the PRD conducted in 2005 and 2006 indicates that the "shop in the front and factory in the back" model, in which there was a clear division of labor between a Hong Kong head office and PRD manufacturing facilities, is gradually becoming less prevalent.[1] Of those firms surveyed, 45.8 percent had no office in Hong Kong, and the ratio of Hong Kong employees to Mainland employees in these companies has steadily declined. In 2002, the ratio was 1 Hong Kong employee for every 120 Mainland employees; in 2006, the ratio was 1 to 170 (HKCER, 2007, 66, 76). After three decades of development, Hong Kong firms are now able to find workers in the PRD to perform tasks that formerly had to be carried out in Hong Kong, and like their domestic Chinese rivals, they will keep employment in Hong Kong to an absolute minimum.[2] In fact, the distinction between Hong Kong and Mainland firms has become quite blurred. Nearly half of the enterprises in the PRD that are Hong Kong-funded enterprises or closely related to Hong Kong are now registered as domestic Mainland firms rather than foreign-invested firms (HKCER, 2007, 110).

Second, there have been changes in the activities that Hong Kong firms undertake on the Mainland. This is partly a result of a slight shift in the competitive advantage of Hong Kong firms. In the 1980s, a Hong Kong firm was able to respond to rising costs in Hong Kong by serving as a middleman: foreign customers were not comfortable operating on the Mainland and Mainland firms were not familiar with the demands of foreign markets. Over time, the need for a Hong Kong intermediary has decreased. Many foreign retailers and brands such as Wal-Mart and Dell have established large and centralized purchasing operations on the Mainland (and often they are not in the PRD), and many Chinese suppliers are comfortable supplying foreign customers directly (HKTDC, 2008b). The response of Hong Kong firms has been to increase the scope of their business activities. According to a survey conducted by the Hong Kong Trade and Development Council (HKTDC), compared with 10 years ago, 67 percent of surveyed companies had increased quality control activities, 62 percent had increased sales and marketing activities, 58 percent had increased product design and development, 51 percent had increased product procurement, and 50 percent had increased corporate social responsibility activities (HKTDC, 2008b).[3] By moving toward a "full package" of services, these firms have increased their value in the global value chain. Hong Kong firms have also been focusing increasingly on new market opportunities within China. The HKTDC survey (2008b) indicated that almost half of surveyed firms believed that they will establish or expand their presence in the Mainland market over the next 3 years, compared with only a quarter that intended to expand their presence in overseas markets. These firms believe that they have a competitive advantage in the Chinese market vis-à-vis foreign firms, and they intend to exploit this advantage to the fullest.

Although these shifts might appear to represent a de-linking of Hong Kong from the manufacturing operations in the PRD, there is little evidence of this. As Figure 8.1 indicates, regardless of whether a

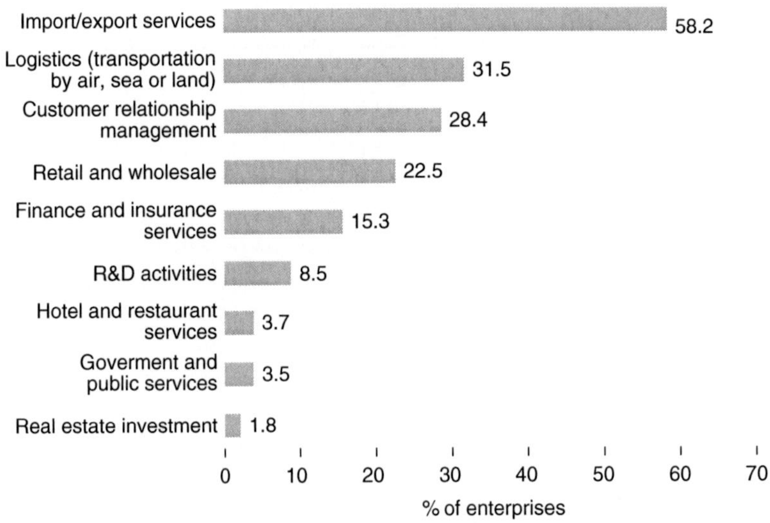

Figure 8.1 Utilization of Hong Kong's service industry
Source: HKCER, 2007, p. 79.

firm had operations in Hong Kong, there is still a strong likelihood of utilizing Hong Kong producer services, particularly those related to import/export services, logistics, and customer relationship management. Demand for Hong Kong producer services is not a function solely of Hong Kong-owned firms utilizing these services, and it is not solely a function of manufacturing operations having a Hong Kong-based head office. Demand is created by the efficiency of these services and the potential savings that can be created for the manufacturing operations that utilize them. The Hong Kong government and the companies that provide producer-related services must continue to ensure that Hong Kong has a competitive advantage in these areas, but this is a familiar challenge, and one that Hong Kong has faced for over a decade.

The most important challenge facing Hong Kong manufacturing firms is not that they will cease to utilize Hong Kong producer-related services as they become more integrated; it is that they will no longer be competitive in the PRD. Costs have been rising dramatically in the PRD over the last 5 years.[4] Firms in the region describe an apparent storm of rising costs. First, wages and the cost of worker benefits have been rising steadily. The minimum wage in Guangdong as a whole increased by an average of 12.9 percent in April 2008; the minimum wage in

Table 8.2 Changes in Guangdong minimum wage

	1 Nov 2002	1 Dec 2004	1 Sep 2006	1 Apr 2008	Change since last increase (%)
Guangzhou	510	684	780	**860**	**10.3**
Zhuhai, Foshan, Dongguan, Zhongshan	450	574	690	**770**	**11.6**
Shantou, Huizhou, Jiangmen	400	494	600	**670**	**11.7**
Shaoguan, Heyuan, Meizhou, Shanwei, Yangjiang, Zhanjiang, Maoming, Zhaoqing, Qingyuan, Chaozhou, Jieyang, Yunfu	360	446	500	**580**	**16.0**
Shenzhen Special Economic Zone	595[a]	690[b]	810[c]	**850[d]/ 1000[e]**	**17.6**

[a]1 May 2002; [b]1 Jul 2005; [c]1 Jul 2006; [d]1 Oct 2007; [e]1 Jul 2008.
Source: *Labour and Social Security Office of Guangdong.*
HKTDC, 2008a.

Shenzhen increased as much as 17.6 percent (see Table 8.2). Factories have been trying to limit the amount of overtime, in order to decrease the wage bill, and this has made it difficult to attract new workers (Interview, 22 July 2008a). Second, the renminbi (RMB) has steadily appreciated. Between June 2005 and July 2008, the RMB appreciated 20 percent against the US dollar and 18 percent against the yen. If an export-oriented firm in the PRD sources 30–50 percent of its inputs (by value) domestically – which survey data suggest is a typical amount – a 10 percent appreciation in the RMB amounts to a 3–5 percent increase in production costs. Finally, energy and commodity prices have been rising rapidly.

A Hong Kong firm in the PRD faced with rising costs has a number of options, but although the overwhelming choice for firms is to upgrade the level of technology and raise the value added of their product, they continue to have limited capacity to do so. In the survey of Hong Kong firms with manufacturing facilities in the PRD conducted by the HKTDC, 22.5 percent of responding firms indicated that they intended to scale down PRD operations, and 3.1 percent planned to shut PRD operations altogether. By comparison, 53.3 percent of firms intended to upgrade technology and increase the value added of their products, and 29.9 percent intended to increase the mechanization of their operations (HKTDC, 2008c). As the manager of Firm A (a Hong Kong electronics firm) ruefully commented, the pressure of rising costs in the PRD is remarkably similar to the pressures that had prompted the firm to relocate production facilities there from Hong Kong in the 1980s (Interview, 18 September 2008b), but this time the response has to be different. If the firm moves to a low-cost site in the interior, it will run into the exact same problem in another 3–5 years.

The problem is that there is little indication that these firms have the capacity to move into higher value-added activities. According to the HKCER survey of firms in Guangdong (see note 1), 65.1 percent of surveyed firms continue to be engaged exclusively in OEM (original equipment manufacturing) activities and only a little over 15 percent were ODM (original design manufacturing) or OBM (original brand manufacturing) (see Table 8.3). Even more telling is that 65 percent of the surveyed firms in Guangdong do not carry out any research and development activities, a strong indication that these firms are engaged in low-technology activities (HKCER, 2007, 95). These are exactly the activities that will face the most difficulties as costs in the region increase. At the end of the 1990s, Firm A could consistently make profit margins of 10 percent in electronic manufacturing services. At this time wages were 600–700 RMB per month (and there was no insurance). In 2008, the profit margins were rarely higher than 3 percent for the same activities. Wages had risen to 1600–2000 RMB per month (plus insurance) and overtime had increased from 1.1 times the normal wage to 1.5 times on a weekday and 2 times on a weekend. The Japanese

Table 8.3 Production mode of Guangdong enterprises

	% of enterprises		
	All Sample	**FIEs**	**OCFs**
Original Equipment Manufacturing (OEM)	65.1	61.9	68.6
Original Design Manufacturing (ODM)	9.8	13.2	6.0
Original Brand Manufacturing (OBM)	6.6	7.0	6.1
OEM & ODM	11.9	10.3	13.8
OEM & OBM	3.0	3.3	2.6
ODM & OBM	1.1	1.4	0.8
OEM & ODM & OBM	2.1	2.4	1.8
Others	0.4	0.5	0.3

Note: FIEs are foreign-invested enterprises (including Hong Kong); OCFs are other contractual firms (which include processing firms and foreign-invested firms that have registered as domestic enterprises).
Source: HKCER, 2007, 56.

electronic gaming company that is the primary customer requests a 2–3 percent price reduction per year (Interview 18 September 2008b).

A related problem is that as the process of industrialization in the PRD broadens and deepens, the industrial profile of the province is shifting away from light industry and the export processing activities that favored Hong Kong firms, toward higher value-added industry. Three of the primary targeted industries in Guangdong, for instance, are electronic information, petrochemicals, and the automotive industries. Shenzhen dominates in the electronic information industry, Guangzhou is dominant in the automotive, petrochemical, and chemicals industries, and both Foshan and Shenzhen play key roles in the manufacture of electrical machinery and special purpose equipment (HKCER, 2007, 37, 40). Between 2001 and 2005, the industry with the fastest growing industrial output value was the automotive industry (see Table 8.4). Increasingly, the light industry will move further inland, where labor

Table 8.4 Industrial output value of nine targeted industries in Hong Kong, 2000–2005

	2000 RMB bn	**2005 RMB bn**	**2001–2005 Average Annual Growth Rate (%)**
Three Emerging Industries	**540.0**	**1,836.3**	**27.7**
Electronic Information	241.8	983.1	32.4
Electrical Machinery and Special Purpose Equipment	162.6	525.7	26.4
Petroleum and Chemical	135.6	327.5	19.3
Three Traditional Industries	**264.4**	**507.3**	**13.9**
Textile and Garments	122.7	215.0	11.9
Food and Beverages	79.9	163.6	15.4
Building Materials	61.8	128.6	15.8
Three High-Potential Industries	**88.1**	**248.6**	**23.1**
Logging and Papermaking	38.8	84.0	16.7
Medicine	18.4	28.7	9.3
Motor Vehicle	31.0	136.0	34.4
Total Industrial Output Value of Nine Industries	**892.5**	**2,592.2**	**23.8**
Total Industrial Output Value of Enterprises above Designated Size	**1,248.1**	**3,594.3**	**23.6**

Source: Guangdong Statistical Yearbooks.
HKCER, 2007, 36.

costs are lower, and this will move them further from the producer services in Hong Kong.

In summary, any public policy with the objective of supporting Hong Kong manufacturing must define this in broad terms. In addition to Hong Kong firms, the entire PRD geographic region must be targeted. This is partly in recognition of the simple fact that it is increasingly difficult to distinguish between Hong Kong and Mainland firms in the PRD, and partly in recognition of the fact that a Mainland firm might be equally as likely to use Hong Kong's producer services as a Hong Kong firm. The policy approach must also be sure to promote capability-building that is relevant to the rapidly evolving industrial structure in Guangdong.

2. Upgrading Hong Kong manufacturing

Hong Kong has its fair share of globally competitive manufacturing firms. Esquel, for instance, is one of the world's leading textile and apparel producers. A high degree of vertical integration – from cotton growing in Xinjiang to spinning, weaving, dyeing, manufacturing, and packaging – enable it to control every step of the production process, and make it a valuable supplier to leading global brands and retailers. The strength of the company lies in the breadth of its international customer base, its understanding of customer needs, and its ability to control every step of the production process in order to satisfy those needs. TTI is a leading producer of consumer and professional products marketed to the home improvement and construction industries. It began as an OEM supplier, moved on to ODM relationships with large US retailers, and then began to purchase global brands. These brands include Milwaukee, AEG and Ryobi power tools and accessories, and Hoover, Dirt Devil and Vax floor care appliances. It has a global design process that integrates the efforts of design centers in the United States, Hong Kong, and the PRD. These are global companies that happen to be based in Hong Kong.

The vast majority of Hong Kong manufacturing companies, however, are in a very different category. As Baark and Sharif (2006, 205) point out, small and medium-sized enterprises (defined as non-manufacturing firms with less than 50 employees and manufacturing firms with less than 100 employees) accounted for 98 percent of the total number of enterprises in Hong Kong in September 2004. These firms are much less likely to undertake design activities or develop brands. As Table 8.3 indicates, the vast majority of Hong Kong firms are engaged exclusively in

OEM activities. These firms are commonly criticized for taking a short-term approach; they have always made money in OEM activities and hence it is difficult to persuade them to make the investments necessary to shift to ODM or OBM activities. One firm, for instance, described how its effort to develop its own brand in the home appliance sector was blocked by its OEM customers. These customers did not want a potential rival, and without the necessary sales channels or the full range of products demanded by a big box store in the United States, the Hong Kong firm was forced to retreat to ODM. Even a company as large and successful as Goldpeak has found it difficult to move into OBM. It does well in Hong Kong and China, but has little hope of developing a brand in Western countries. Batteries are essentially commodities, and the core competency of Western battery brands is marketing and branding.

The Hong Kong government is keenly aware of the challenges that face these local manufacturing firms, and as part of the broader effort to support the development of high-technology industries in Hong Kong, it has been engaged in promoting the upgrading efforts of these firms. At the core of these efforts are the five R&D facilities established in 2005, 2006, two of which focus on mid-range technologies for traditional manufacturing firms: the Automotive Parts and Accessory Systems Center (APAS) hosted by the Hong Kong Productivity Council (HKPC), and the Hong Kong Research Institute of Textiles and Apparel (HKRITA) hosted by the Hong Kong Polytechnic University. Each of these centers was initially given funding for a 5-year period (APAS had HK$350 million and HKRITA had HK$275 million). Their objective is to increase the competitiveness of Hong Kong industry by developing and transferring technology and designs to Hong Kong firms, by supporting and developing the human capital that these industries require, and by providing consulting and market intelligence activities to firms.

How can an R&D facility support the development of SMEs in Hong Kong? For the sake of simplicity, the challenges that Hong Kong manufacturing firms face can be grouped into two categories: a "technology gap" and a "marketing gap." As Hubert Schmitz explains, the technology gap is a result of being removed from international sources of technology (and in particular the feedback loop between users and producers that spurs innovation), the difficulty of accessing proprietary technology, and weak national and/or local support for innovation (Schmitz, 2007, p. 420). These technologies may include the "hard" technologies that are embodied in machinery, or "soft" management systems such as quality control or supply chain management. The marketing gap is a result of the difficulty an export-oriented firm will have understanding and responding to rapidly changing consumer demand in foreign markets. It is exacerbated by highly concentrated retail sectors in these markets (which shifts leverage within the value chain to the buyer) and the capital intensity of developing a brand.

2.1. Textiles and Apparel

The objective of HKRITA is to assist Hong Kong textile and garment firms in closing the technology gap. The intent is to leverage the collective R&D of universities in Hong Kong, the Mainland, and overseas to strengthen the capabilities of the Hong Kong textile and apparel industry.

The size of this industry makes it an obvious target: it accounts for 27.8 percent of employment in the Hong Kong manufacturing sector (46,000 workers). As the *Made by Hong Kong* study argues, Hong Kong textile and apparel firms have many advantages – highly flexible production systems, long experience and good connections with foreign customers and markets, and a strong ability to coordinate widely dispersed production networks (Berger and Lester, 1997, chapter 9). One element of upgrading involves bolstering these traditional advantages. This set of issues falls primarily under the purview of the logistics and supply chain management R&D center, the subject of another chapter in this volume. A second component of upgrading involves the development of new fabrics and processes. Although the industry might appear to be quite traditional, new materials (such as breathable fabrics, wrinkle-free fabrics, stain-resistant fabrics, anti-bacterial fabrics, and anti-UV fabrics) are at the core of innovation in the sector, and it is in this area that HKRITA has focused its efforts. Its core research areas are new materials and textile and apparel products, advanced textiles and clothing production technologies (such as new coloration, finishing, and spinning technologies), product design and evaluation technologies, and enhanced industrial systems and infrastructure (Interview, 14 March 2008d).

As is the case with the other R&D centers, the primary purpose of HKRITA is to coordinate the relationships between firms and research organizations. Hong Kong Polytechnic University, the host institution for HKRITA, undertakes the research and it has strong capabilities in textile and apparel production. Firms are able to sponsor research in three ways. In a platform project, firms pay a total of 10 percent of the project (so any one firm will be paying less than 10 percent) and the government pays the remainder. HKRITA controls the rights to the intellectual property (IP) and participating firms receive a discount when they license the technology. In a collaborative project, a single firm pays less than 50 percent of the cost and it has the exclusive right to license the technology (but HKRITA owns the IP). In an exclusive project, a firm pays 51 percent or more of the project and it owns the IP.

A large Hong Kong firm in this sector will usually have significant internal R&D capability. Firm B, for example, is a leading firm in the industry and it has 40 staff in its R&D facility in the PRD, with backgrounds in textile chemistry, textile engineering, and engineering (for waste water treatment). Internal R&D is quite practical and applied, however, and the firm also has a strong need for more basic science research – processes like dyeing and finishing are essentially chemical ones. For these projects the firm relies on collaborations with both Hong Kong universities (coordinated by HKRITA) and Mainland universities (such as Zhejiang Science and Technology, Wuhan University, and Shanghai Donghua). Each university has particular strengths, but overall the skills of Hong Kong and Mainland universities are comparable, according to the head of the firm's R&D department (Interview, 20 July 2008). Although Hong Kong Polytechnic is excellent, Shanghai Donghua (formerly named China Textile University), for instance, has been conducting research in the field since the 1950s, and is a key institution in China for the study of textile engineering, material science, textile chemistry, and dyeing and finishing engineering. It has key national laboratories and engineering research centers in these fields and a science park.

The primary problem for Hong Kong research projects is the high cost and the difficulties of transferring technology to firms. Firm B has undertaken many projects with Hong Kong Polytechnic, but they have always been either platform or collaborative projects, so the firm has never owned the IP. When it has tried to buy or license technology it has found the process to be long and expensive (although this might be improved by HKRITA – it is still too soon to say). When Firm B works with Mainland institutions, it pays the full cost of the project (because the cost is much lower) and as a result it controls 100 percent of the IP (Interview, 20 July 2008). Although IP protection is a potential problem, thus far they have been able to structure the contracts with Mainland institutions so that the protection of IP rights is not a major issue.

In short, collaborating with Mainland universities is cheaper and in many respects more convenient. Although an obvious solution would be to allow HKRITA to work directly with Mainland universities as well as with those in Hong Kong, government regulations do not allow this. A Hong Kong university is able to hire a Mainland university as a consultant on a project, but funding cannot go directly from HKRITA to a Mainland university. There are two reasons for this. First, the government is interested in promoting research activity in Hong Kong. Second, the government is concerned that it will not be able to audit sufficiently the use of Hong Kong funds on the Mainland (Interview, 28 October 2008).

Firm B is a large and globally competitive firm, and most of the clients of HKRITA tend to be large firms (Interview, 14 March 2008d). The absence of SME clients points to an additional problem: the high cost of research in Hong Kong has the potential to prevent the center from supporting those firms that need its assistance the most. These smaller firms – the SMEs that have no R&D capability of their own – are most likely reluctant to pay the fees that are required by Hong Kong universities. In the HKCER survey of firms in the PRD, 80 percent of the surveyed enterprises considered high cost to be the primary problem of conducting R&D activity in Hong Kong (HKCER, 2007, 98). While supporting the development of local research capabilities is a worthy goal, the Hong Kong government may have to make a decision as to whether its primary policy goal is to ensure that funds are spent within Hong Kong institutions, or whether it should seek to support development among firms.

2.2. Automotive

If the textile and garment industry is an obvious choice for government support in Hong Kong, the automotive industry is rather less so, since Hong Kong does not have much of an automotive industry. There are approximately 280 firms in Hong Kong that are auto-related, and most of these produce aftermarket parts for export – the lowest category of activity in the automotive value chain.

Although it might seem strange to create an R&D center for an industry that does not really exist, it begins to make more sense when one considers that the car is increasingly an electronic product. Approximately 35 percent of the value of a car is currently in its electronics and this percentage is increasing. According to the Hong Kong Productivity Council (HKPC), in the next 3–5 years, 90 percent of the innovation in the sector will be in onboard electronics (e.g., collision control, navigation systems, voice recognition, traction control, tire pressure monitors, etc.).[5] The car is rapidly becoming more of an electronic product than a mechanical one, and unlike autos, electronics is a dominant industry among Hong Kong manufacturing firms. Perhaps even more importantly, the automotive industry is the fastest growing industry in Guangdong (see Table 8.4). Ideally, firms would move out of OEM activities in electronics – a sector that is rapidly losing comparative advantage in the PRD – and into a sector that is growing rapidly in the PRD. The sales value of auto electronics in China in 2007 reached RMB 86.76 billion (US$12.2 billion), an increase of 40 percent over the previous year, and is expected to reach RMB 240 billion by 2011.[6] Tire pressure monitoring systems, for example, became mandatory in all new cars sold in North America in 2007, but are included in only 5 percent of Chinese vehicles.[7]

The key question is whether Hong Kong firms will be able to develop the capabilities needed to compete in the automotive industry. The development literature gives reason for pessimism. The "technology gap" referred to earlier is particularly high in the auto sector because the industry is dominated by the assemblers and the top tier of global suppliers. Design costs in the industry are extremely high, and these high costs create strong incentives for the global assemblers to (1) capture global economies of scale (in order to spread the cost of design over larger volumes, and (2) push part of the burden of design onto the Tier 1 supply firms. The latter dynamic serves to limit the range of opportunities for small suppliers in developing countries because Tier 1 suppliers must follow the global assemblers to new production sites (Humphrey and Memedovic, 2003). The Toyota supply network in Guangdong, for instance, shows strong evidence of this "follow-sourcing"; overwhelmingly, the key suppliers are Japanese firms that Toyota has brought to China (see Table 8.5). These firms will use local Chinese suppliers in lower tiers of the supply chain, but these tend to be low value-added activities.

Will global firms dominate in China? There is an important reason why the conventional wisdom might not apply in this case. Discussions of industrial development in East Asia generally focus on export-led growth, and this focus on export markets defines the nature of the challenge that home country firms face. The marketing gap is large because firms are far removed from final markets; the technology gap is large because the objective is to reach the cutting-edge technical and quality standards demanded by foreign customers. China presents an unusual opportunity because the focus is on the domestic market. Domestic demand for passenger vehicles has increased from 504,562 units in 1998 to 6.3 million vehicles in 2007, and China is now the second largest market for passenger cars in the world. Within this domestic market there

Table 8.5 Ownership of Toyota Group Suppliers established in Guangzhou in 2004

Firm	Investment Share	Major Products	Major Customers
Fengai Guangzhou Automotive Seat Parts	Toyota Boshoku 51%; Aisin Seiki 49%	Seat backs, cushions, tracks, etc.	Guangzhou Intex Parts
Guangzhou Intex Auto Parts	Toyota Boshoku 50%; Takanichi 25%; GAIC Auto Parts 25%	Seats, door trim, headliners, etc.	Fengai Automotive Seat Parts
Toyota (Foshan) Gosei Auto Parts	Toyota Gosei 65%; Taiwan subsidiary 30%; Toyota Tsusho 5%	Interior and exterior parts	Toyota and other Japanese automakers
Toyo Automotive Parts (Guangzhou)	Toyo Tire & Rubber 100%	Rubber NVH products	Japanese automakers
Aisan (Foshan) Autoparts	Aisan Industry 95%; Yoyoda Tsusho 5%	Fuel injectors, related engine parts	Toyota and other Japanese automakers
Foshan Tokai Rika Automotive Parts	Tokai Rika 100%	Key-lock sets, seat belts	Toyota and other Japanese automakers
Aisin Seiki Foshan Automotive Parts	Aisin Seikie 100%	Engine parts	Toyota and other Japanese automakers
Huizhou Zhucheng Wiring Systems	Sumitomo Wiring 20.4%, Shenzhen Dongf. 25%, Shenzhen Sumitomo Equipment 24%	Automotive wire harness	Japanese automakers
Guangzhou Hayashi Telempu	Hayashi Telempu 51%, GAIC 49%	Interior parts	Japanese automakers
Mitsui Chemicals Plastic Compounds (Zhongshan)	Mitsui Chemicals 100%	Polypropylene compounds	Japanese automakers

Source: "Toyota suppliers advance in Guangzhou," *Fourin China Auto Weekly*, 27 December 2004, p. 2.

are distinct segments that have a variety of demands for quality and performance. The fastest-growing component of the market over the last 10 years has been individual first-time car buyers. In 1995, 25 percent of Chinese automobiles were registered to private individuals. In 2005, this figure stood at 55.3 percent (14.97 million units) and it has been rising steadily since.[8] Consumer research in this segment indicates that more females are buying cars, and consumers are increasingly well educated and wealthy. These consumers demand high value for money, a variety of styles, and are more likely to be swayed by the recommendation of a friend or relative than by the particular make or model of car.[9] Between 2001 and 2007, the number of models of small car sold in China increased from 4 to 19 and the average price decreased by nearly 33 percent (from US$10,566 in 2001 to US$6,931 in 2007; Mei, 2007).

Rapid growth at the low end of the market and intense price competition have created opportunities for domestic Chinese auto assembly firms. In some respects, the products that the multinationals bring to China are too advanced and over-engineered for the lower segments of the Chinese market, and because the objective of these firms is to maximize global economies of scale, they are reluctant to create designs specifically for the Chinese market. This has created an opening for independent Chinese firms that use low-cost suppliers, less capital-intensive manufacturing techniques, and simpler designs than their foreign competitors. At the end of the 1990s, this category of firm occupied a small fraction of the marketplace, but by 2007 they controlled a third of the market.[10] Anhui-based Chery, the most successful of these firms, began producing cars in 1999 and only 4 years later it had achieved annual sales of 80,000 vehicles. In March 2008, Chery sold more vehicles in China than any other manufacturer.

In short, the low end of the market provides a critical learning opportunity for indigenous firms and offers the potential for them to increase their operations. Because foreign firms are rarely able to meet the price demands of consumers in this segment of the market, the domestic firms have the benefit of being insulated from foreign competitors. The high end of the market remains the domain of the foreign firms, and the Chinese firms rarely have the expertise to design, manufacture and market products that can compete in this segment of the market. Neither foreign nor domestic firms are content to stay in their respective segments, however. The domestic firms want to upgrade and escape the intense competition at the bottom of the market; the foreign firms want to lower costs so that they can compete in the rapidly growing middle segments of the market (Brandt and Thun, 2009).

This segmentation of the domestic market in China, and the competition between domestic Chinese and foreign firms, creates an opportunity for Hong Kong firms seeking to develop new capabilities. On the one hand, a global assembler will continue to use global Tier 1 suppliers, but due to the intense price pressure, these global suppliers will make every effort to utilize lower-cost suppliers. The global Tier 1 firm works with the low-cost supplier to improve and maintain quality levels, but also takes care to ensure that it is only outsourcing the lower value-added activities. On the other hand, the domestic Chinese assemblers, because they have limited R&D capabilities, will give less assistance to their suppliers, but they are eager to allow a supplier to engage in a full range of design activities.

The technical director at one Chinese supply firm compared the relationships with a foreign versus domestic assembly firm to a rectangle that is sitting on end as opposed to one that is lying flat. The former symbolizes the relationship with a foreign company that is seeking to lower its costs: it is narrow and deep. The domestic supplier can achieve a high level of competence very quickly because a global supplier will be assisting them, but the range of capabilities will be narrow. The latter represents the relationships with a domestic assembler that seeks to upgrade: the domestic supplier has an advantage over foreign suppliers on cost and it can engage in a wider array of activities because the technical demands are lower, but it learns less because the domestic assembler is not in a position to provide as much assistance. The objective of a Hong Kong firm should be to maximize the benefits of participation in multiple value chains.

APAS is in a position to provide critical assistance to Hong Kong firms seeking to develop automotive components. The research center is wisely concentrating its efforts on automotive systems that maximize the experience that Hong Kong firms have in electronics, and it is trying to develop expertise with products that are not too sophisticated (because it will not be able to compete with global firms), but are more sophisticated than the average Mainland supplier will be able to handle. Rather than focusing on airbags and anti-lock braking systems (i.e., core systems), for instance, the center focuses on collision alert, GPS, adaptive headlight systems, and audio/visual units (Interview, 14 March 2008b). The center has three areas of focus – advanced materials, safety, and software/electronics – and it provides a range of services to firms. First, it can prove the testing that is necessary to achieve the certifications and international standards that are critical in the auto component sector, and it can provide the analysis that will allow firms to achieve these standards. Second, it is able to provide SMEs with turnkey solutions or testing in areas that require expensive machinery (such as the machining necessary for molds) that a SME might be reluctant to invest in. Third, it coordinates projects and serves as technical consultant. Firms have the same three options that they have with HKRITA (a platform project, collaborative project, or exclusive project), and often APAS plays the role of matchmaker, bringing together firms that have the complementary skills needed for a new project.

Within the HKPC, the host institution of APAS, firms are able to find people with the specialized technical skills needed for a project. A headlight project, for instance, involves three to four optical consultants, four people to create the algorithms for the required software, and two to three people in mechanical design (Interview, 14 March 2008b). As the design manager in one Hong Kong firm explained, firms would probably be able to create these capabilities themselves, but it would take a great deal longer and they would have to generate the capital to support the project. Firms are skilled in manufacturing and the electronics industry, but have little experience with automotive technology. The problems that the firms encounter are high-level technical problems, and these are exactly the type of problems that the specialists at HKPC are able to provide assistance with (Interview, 18 July 2008b). The intent is not necessarily to find a better solution than a major global player such as Toyota would be able to provide, but to develop a low-cost and innovative solution that could then be marketed to domestic firms in China or foreign firms that are struggling to lower costs.

The primary difficulty, and the primary area in which firms could use more help from APAS, is with finding customers. Most of the Hong Kong firms do not have connections in the industry, and it is difficult to establish relationships with the assembly firms. Beyond the obvious need for a buyer for the product that is being developed, the lack of a customer can slow development time. Often the products are not plug-and-play, but require integration with the specifications of a particular vehicle, and the software cannot be completed until the supplier has the specs from the customer.

3. Conclusion

The challenges that Hong Kong manufacturing firms face are in many respects the classic problem of industrial development. As the development process proceeds, costs will rise and firms must either move to a low-cost region or move into higher value-added activities. When

Hong Kong firms faced the same set of challenges two decades ago, they chose the former option. This time, many hope to pursue the latter option.

Although the challenges of upgrading are commonplace, the context of "one country two systems" is not, and Hong Kong must seek to maximize the advantages of its relationship with China, while minimizing the disadvantages. One primary advantage is the opportunity created by the Chinese domestic marketplace. The East Asian "model" of economic development has traditionally been one of export-led growth. It is linkages with global value chains that provide the opportunity for upgrading because knowledge flows through the chain from global buyers to local suppliers (Humphrey and Schmitz, 2002, 1020); industrial development is an iterative process of "learning-by-doing," as global buyers work with suppliers in order to assure quality standards, service, and performance (Gereffi, 1999). As I have argued in this chapter, the focus on export markets accentuates the size of the technology and the marketing gaps faced by local firms: in order to play a leading role in a global value chain, a firm must compete with the best in the world.

While many Hong Kong firms are leading competitors in the global marketplace and many others should be striving to compete globally, there is a large group of manufacturing firms that are stuck in the labour-intensive low value-added activities of global value chains. For this latter category, the domestic market in China offers an opportunity because it provides a lower rung on the upgrade ladder. The technical demands of Chinese buyers are not as high and preferences may vary from global markets. Hong Kong firms are in a position to recognize and meet the market demands of their "one country" more flexibly than global firms, and this will allow them to increase production volumes and broaden the scope of their activities. Hong Kong firms should not be content to remain in the low-end segments of the Chinese market, however; the intent should be to use these lower rungs on the upgrade ladder to gain the capabilities needed to compete in global markets.

In order to support the upgrading efforts of firms, the Hong Kong government must seek to minimize the disadvantages of the "two systems" and continue to move toward a regional approach to development. At firm level, the objective of government policy should be to promote industrial upgrading and economic growth in the PRD region as a whole, rather than just in Hong Kong-owned firms, both because firm ownership does not determine a firm's likelihood to utilize the producer services of Hong Kong and because it is increasingly difficult to distinguish between Hong Kong and Mainland firms based on ownership categories.

There are indications that this redefinition is well under way. APAS, for instance, is able to work with Guangdong as well as with Hong Kong firms. The government can support these efforts by working at governmental level to coordinate R&D efforts in Guangdong with similar efforts in Hong Kong – these should be complementary rather than competing efforts. An important benefit of coordinating efforts in the auto industry is that government involvement on the Mainland will help secure customers for the firms that APAS is working with on projects. The municipality of Guangzhou is a joint venture partner with Honda, and Guangzhou auto has aspirations to develop an independent brand. This is exactly the sort of firm that APAS should be seeking to support. A more regional approach facilitates the ability of Hong Kong firms to access the Mainland market.

The R&D efforts that have been created to support industrial upgrading in general manufacturing should also be allowed to break free of a parochial Hong Kong perspective. Rather than hindering the efforts of the Hong Kong R&D centers to deploy the resources of low-cost research in Mainland universities on behalf of SMEs, the Hong Kong government should work to make collaborations with Mainland universities more effective. The research activities funded by the R&D centers are unlikely to support the development of core capabilities in Hong Kong universities and unlikely to solve the funding problems that David Mowery highlights in this volume; the research is generally small scale and seldom employs cutting-edge technologies. As a result, it makes more sense to focus on company needs, and the primary need is lowering the cost of R&D support. If firms are allowed to use government research funds in both Hong Kong and the Mainland, they will be able to reap the advantages of both systems.

The Hong Kong government is unlikely to adopt a muscular industrial policy to support the upgrading of manufacturing firms, and it probably should not do so. Firms that manufacture in the Pearl River Delta play a vital role in supporting Hong Kong's producer-related services, but they are a relatively small part of Hong Kong's economy. The modest role that the government is carving out for itself makes sense. It can play an important role in coordinating projects and providing services that SMEs are unlikely (or unable) to provide for themselves, and, in doing so, it can promote the economic health of the region upon which it depends. It can continue its efforts to improve the functioning of the "two systems" that coexist within one country. In these small but critical ways, the government can help firms create a sustainable manufacturing base in the Pearl River Delta, a base that will have Hong Kong at its core.

Notes

1. The HKCER survey was conducted between 20 September 2005 and 10 March 2006. Surveys of firms that responded to initial contacts were conducted by surveyors based on a questionnaire. There were ultimately 2529 valid questionnaires. For further details on the survey see HKCER, 2007, 45.
2. One of the surprising results of the HKCER survey in 2005–2006, for instance, is that fewer surveyed firms were conducting financial management in Hong Kong (27.8 percent) than in Guangdong (31.4 percent). 30–40 percent of surveyed firms assigned the responsibilities for financial management, IT management, R&D and logistics equally between their Hong Kong offices and PRD factories (2007, 69).
3. In this survey a total of 2230 valid replies were received. 38 percent were from traders and 62 percent from manufacturers-cum-traders. About two-thirds of respondents had engaged in trading business for 10 years or more. As is true of Hong Kong trading companies overall, most of the surveyed firms were SMEs: 40 percent of the respondents handled goods worth less than HK$10 million, 50 percent between HK$10 million and HK$100 million, and the remaining 10 percent over HK$100 million.
4. The data in this paragraph is from HKTDC, 2008a.
5. HKPC, "Advanced Automotive Electronics," Presentation, February 2008, p. 13.
6. "China's Auto Electronics Sales Jump 40% in 2007," *Asia Pulse*, 13 March 2008.
7. Megan Lampinen, "China: Lear to launch production of tyre pressure monitoring systems," *Automotive World*, 25 March 2008.
8. "Vehicle Ownership in China" *Fourin China Auto Weekly*, 5 December 2005.
9. Surveys of purchasing decisions by small car owners indicate that between 2001 and 2007 the percentage of people who bought primarily on the basis of brand reputation decreased from 10 to 5 percent, the percentage were simply seeking a particular model decreased from 13 to 10 percent, and the percentage that were swayed by the recommendation of a relative or friend increased from 4 to 8 percent (Mei, 2007).
10. "China's Independent Automakers: Independent Brands Climbed to No. 2 Spot in 2005," *Fourin China Auto Weekly*, 8 May 2006; "Carmakers brace for touch small car race," *China Daily*, 22 April 2008.

References

Baark, E. and N. Sharif (2006) "From trade hub to innovation hub: The role of Hong Kong's innovation system in linking China to global markets." *Innovation: Management, Policy & Practice* 8(1–2): 193–209.

Berger, S. and R. K. Lester, eds. (1997) *Made by Hong Kong*. Hong Kong, Oxford University Press.

Brandt, L. and E. Thun (2009) The fight for the middle: Upgrading, competition, and industrial development in China." *Sloan Industry Studies Working Paper WP-2009-4.*

Enright, M. J., and E. A. Scott (2007) The Greater Pearl River Delta. Hong Kong, Invest Hong Kong.

Gereffi, G. (1999) "International trade and industrial upgrading in the apparel commodity chain." *Journal of International Economics* 48: 37–70.

HKCER (2007) Made in PRD: Challenges & opportunities for HK industry. Hong Kong, Hong Kong Centre for Economic Research commissioned by the Federation of Hong Kong Industries.

HKTDC (2008a) "Upward pressure on export prices mounting but competitiveness maintained." 17 September.

HKTDC (2008b) "The changing business model of trading companies in Hong Kong."

HKTDC (2008c) "Beyond cheap labour: Building a competitive edge through adding value." 5 September.

Hong Kong Policy Research Institute (2003) Neo-industrialization in Hong Kong: finding a place for Hong Kong industry, and rising to the global challenge. Hong Kong, Hong Kong Policy Research Institute.

Humphrey, J. and O. Memedovic (2003) The global automotive industry value chain: What prospects for upgrading by developing countries. Vienna, United Nations Industrial Development Organization, 1–50.

Humphrey, J. and H. Schmitz (2002) "How does insertion in global value chains affect upgrading in industrial clusters?" *Regional Studies* 36(9): 1017–1027.

Mei, S. (2007) "Shifting dynamics of china's small car segment."

Schmitz, H. (2007) "Reducing complexity in the industrial policy debate." *Development Policy Review* 25(4): 417–428.

Tao, Z. and R. Y. C. Wong (2002) "Hong Kong: From an industrialised city to a centre of manufacturing-related services." *Urban Studies* 39(12): 2345–2358.

9

Biotechnology in Hong Kong: Prospects and Challenges

Joseph Wong
University of Toronto

1. Biotechnology and uncertainty

The global biotechnology revolution purportedly began during the late 1970s, when the biological "heuristic" in health care technology was expected to both rival and ultimately prove superior to existing chemistry-based approaches to health and health care. Rooted in new discoveries in genetics and the promise of genetic engineering, and fuelled by a flurry of government research support, venture capital and increasingly entrepreneurial universities, biotechnology was expected to revolutionize how human therapeutics were developed, screened, and delivered. Biotechnological tools were expected to rationalize drug development. They would lead to new diagnostic tools. Recombinant DNA techniques would allow scientists to re-engineer cells to produce new and "smarter" proteins, the basis for a new generation of therapeutics. The introduction of biotech would in effect restructure the global human health care industry, as pharmaceutical firms increasingly turned to smaller, specialized and more cutting-edge biotech firms for new screening techniques and drug candidates. Biotechnology was imagined as an enabling technology, a platform technology, and a source of knowledge for advancing human health care. Simply put, the possibilities for applying biotechnology to health and health care seemed endless.

As with all revolutionary moments, however, the growth of biotechnology and the future development of a global biotech industry were inherently uncertain. As a science-based industry, it was unclear whether the science of biotechnology would actually result in more efficient and more effective health care interventions: will it work? Venture capital's enthusiasm notwithstanding, there was tremendous uncertainty surrounding biotech's economic viability as well, especially seeing as new discoveries at the time were still considerably far-from-market: will it have value (Pisano, 2006)?

Between 1998 and 2003, US$85 billion was invested by the private sector in the US. Nearly US$30 billion is expended by the US government each year for upstream life sciences research (Casper, 2007). Yet, despite such large-scale investments, the global (and US) biotechnology sector, as an industry, has fallen far short of initial, albeit uncertain, expectations. Recent data from the 2000s, for instance, show that biotechnological techniques have not in fact resulted in more effective drug development processes. Biopharmaceuticals have not revolutionized the field of human therapeutics, nor have they radically altered the business model of the conventional pharmaceutical industry. And economically, the global biotechnology industry has not fared particularly well (Hopkins et al., 2007). While there have been some – a small few – major success cases, the industry as a whole has lost billions of dollars (US$40 billion according to a 2004 *Wall Street Journal* report). Two firms, US-based Amgen and Genentech, account for nearly 50 percent of all positive cash flow in the biotech sector. As of 2003, there were nearly 1500 biotechnology firms in the US alone, though less than 200 biotech products had actually made it to market (Pfeffer, 2005 104).

It was against this uncertain backdrop that Hong Kong entered into the biotechnology sector during the early 1990s. Driven by the government's new industrial upgrading initiatives, and by leveraging Hong Kong's entrepreneurial spirit and transparent corporate regulatory environment, Hong Kong looked to make significant inroads into life sciences industry (Berger and Lester, 1997). Government commitment was high, by Hong Kong standards. Public funding for biotechnology R&D, allocated through the government's industry support fund (ISF), increased from just 7 percent in 1994/1995 of the fund's total investment to over 40 percent in 1997/1998 (Tsang and Lo, 1998, 151). The Hong Kong Institute of Biotechnology (HKIB) was founded during the late 1980s, and an on-site incubation center for start-up firms was formed in 1996. The Biotechnology Research Institute (BRI) was established at the Hong Kong University of Science and Technology (HKUST) in 1990. In 2001, the Hong Kong Jockey Club Institute of Chinese Medicine was formed. The Chief Executive Tung Chee-Hwa announced during the late 1990s the government's 10-year blueprint for the modernization of Hong Kong's traditional Chinese medicine (TCM) industry. Simply put, a sector had been put into motion.

Though significant efforts were made to launch a domestic biotech industry, a sense of uncertainty regarding biotechnology's commercial prospects nonetheless quickly prevailed in Hong Kong. Reality set in. Technological and economic uncertainty in the global biotech sector is exacerbated by the fact that Hong Kong is a late entrant into the life sciences sector. Hong Kong is a laggard in technological innovation more generally. Research and development (R&D) spending amounted to just 0.38 percent of GDP in 1995 and had increased to only 0.79 percent in 2005. Meanwhile, other East Asian competitors such as Singapore and Taiwan spend 2.5 percent of GDP for R&D; Korea leads the way among late developing economies in the region, expending just under 3 percent in 2005. Moreover, in Hong Kong, government resources account for a relatively large portion of R&D expenditures. Industry makes up approximately 30 percent of the total R&D spending (Baark, 2005, 8), whereas in most other advanced economies firms account for between two-thirds and three-quarters of the national R&D bill. In the field of biotech specifically, the number of firms in Hong Kong is quite small. The government suggests that there are between 250 and 300 "biotechnology-related companies," though this figure represents a rather expansive definition of biotechnology.[1] Industry insiders estimate that Hong Kong is actually home to less than 50 "true" biotech firms. Indeed, because of reasons related to scale, or lack thereof in Hong Kong, biotechnology's and bioindustry's uncertainties are magnified intensely. With a population of just 7 million people, Hong Kong's efforts to become a cutting-edge technology innovator are continually frustrated due to a relatively small talent pool, fewer resources in general, and the absence of a critical mass of firms in the life sciences sector. This chapter illuminates ways in which Hong Kong may begin to overcome some of these challenges.

2. Biotechnology in Hong Kong

Despite such uncertainty – after all, technological and economic uncertainty surrounding the biotechnology sector is a global concern – Hong Kong has made significant strides in building up its R&D capacity in the life sciences. In the biotechnology field specifically, almost all R&D funding comes from government coffers. As I indicated above, the ISF allocated over 40 percent of its funds to biotechnology projects during the mid to late 1990s. Prior to 1998, the applied research fund (ARF), the government's investment fund earmarked for industry, allocated HK$16.6 million of its total HK$97.3 million, or 17 percent, to biotechnology firms, of which one, Hong Kong Transgenic Ltd, was in fact an equity investment rather than a loan. By 2005, the government's Innovation and Technology Fund (ITF), which had by then subsumed the ISF, had supported 57 biotechnology projects, accounting for 9.2 percent of the total number of ITF-sponsored R&D projects. These projects were funded by about HK$161 million, or nearly 10 percent of all ITF funds. Combined with R&D resources allocated for traditional Chinese medicine (TCM), life sciences-related industries accounted for over 13 percent of the total ITF disbursements through 2005.

The government's appetite for the uncertainties of the biotechnology industry began to wane by the early to mid 2000s. After consultations with stakeholders in industry, government and the research community, the Innovation and Technology Commission (ITC), which was established by the government in 2000, unveiled in 2004 its report *New Strategy of Innovation and Technology Development*. The ITC identified 13 specific technology focus areas in the report. Biotechnology was not included. The ITC also introduced a new three-tiered funding system for R&D. Regular funds were to be provided for Tier 1 and 2 technologies. Biotech, however, was reclassified as a Tier 3 technology, which meant that while it remained a government priority, funding for biotech was to be on an "exceptional basis" (Wan, 2005, 919). Five new R&D centers were established in 2006 to promote applied R&D. Again, biotechnology was excluded. The proportion of ITF-sponsored projects (counted as number of projects) in biotech continued to hover at around 9 percent through 2008. However, while the absolute amount of R&D funds for biotech remained at pre-2005 levels, the proportion of R&D funds (counted in dollars) granted to biotech projects decreased from almost 10 percent of all ITF funds in 2005 to just 6.8 percent in 2008. Moreover, the ARF, which had invested about 17 percent of its funds in biotech firms prior to 1998, ceased to make any new investments after 2004; during the intervening years 1998–2004, the ARF invested in only one additional biotechnology firm, Plasmagene Bioscience Ltd, an investment which amounted to HK$11.7 million or less than 3 percent of the ARF's investment portfolio between 1998 and 2004 (ITC website).

In addition to new injections of R&D funds designated for life sciences research during the 1990s and into the 2000s, significant efforts were made to create new dedicated research institutions in order to strengthen Hong Kong's R&D capacity in biotechnology. With an endowment of HK$300 million from the Hong Kong Jockey Club, the Hong Kong Institute of Biotechnology (HKIB) was established in 1988. Envisioned initially as a center of applied biotech R&D, the HKIB was refashioned during the mid 1990s into an incubator for start-up biotechnology firms. The Biotechnology Research Institute (BRI) was established at the HKUST in 1990 with an endowment of HK$130 million, again from the Hong Kong Jockey Club. The Genome Center was created in 2002 as part of Hong Kong University's Li Ka Shing Medical School with an initial budget of HK$120 million. The Hong Kong Jockey Club Institute of Chinese Medicine (HKJCICM) was founded in 2001 as a subsidiary of the Applied Science and Technology Research Institute (ASTRI) with a HK$500 million donation from the Jockey Club. Functioning as both an R&D facility and a research funding agency, the HKJCICM endeavored to become Hong Kong's premier center for the "modernization" of the TCM sector. Parallel efforts were made at the Chinese University of Hong Kong (CUHK) where, in 2000, its longstanding Chinese Medicinal Material Research Center was expanded to form the CUHK Institute of Chinese Medicine (ICM).

Eclipsing all these efforts to broaden institutionally Hong Kong's R&D base in the life sciences is the recent construction of the Hong Kong Science Park (HKSP), located adjacent to the CUHK, near the Sha Tin industrial park and under an hour's distance from Shenzhen, Guangdong Province by train or bus. Phase 2 of the Park, completed in 2008, includes two buildings dedicated to biotech R&D. A total of 14 floors of laboratory and office space have been made available to local and foreign life sciences firms. In order to attract start-up firms, the Science Park authorities have kitted out an entire floor (with more in the plan as needed) in the biotech R&D buildings with basic laboratory benches and communal facilities. The Park also features an "enhanced" biotech incubation program for start-up enterprises, providing pre-venture firms with not only lab space and a biotech cluster, but also support for important services in intellectual property (IP) management, investor relations, and legal advice. The HKSP has recruited a new leadership team comprising Hong Kong returnees who bring with them bioindustry experience from abroad.

There is talent in Hong Kong. The consensus among local and foreign analysts of the life sciences sector is that Hong Kong's upstream research capacity is quite strong and internationally competitive. Unlike elsewhere (East Asia, Europe or North America), the majority of R&D funds in Hong Kong is allocated for research conducted within universities. Higher education R&D accounts for two-thirds of all R&D spending in Hong Kong. Post-secondary student enrollment in the sciences, specifically in the life and medical sciences, has been consistent and very high (HKSTP biotechnology initiative internal report, 2004). The major universities have also been transformed into high-performing R&D centers.

Not surprisingly, upstream research output among Hong Kong's tertiary institutions is impressive. Erik Baark (2005, 10) shows that science and engineering faculty at Hong Kong universities account for the majority of all research publications (the remainder being in the humanities and social sciences). In 2001/2002, of the total of 26,996 research publications from Hong Kong's eight leading universities, 15,602 or 58 percent were from science and engineering faculty. Of that, biology and medicine accounted for the largest share with 6,529 publications, more than the engineering field. In terms of quality of research, studies have shown that, beginning in the late 1990s, the impact factor for Hong Kong's international academic publications in the life sciences field, while falling short of leaders such as the UK, nonetheless ranks among several northern European countries. The vast majority of these high-impact publications have been concentrated in Hong Kong's two medical universities, the University of Hong Kong (HKU) and Chinese University of Hong Kong (CUHK), as well as Hong Kong University of Science and Technology (HKUST) (HKSTP biotechnology initiative internal report, 2004).

Recognizing that a solid basis in upstream research can be utilized further downstream, technology licensing and transfer offices were established inside research-oriented universities. Two of the largest and most active, in terms of disclosures and patents, are located at the HKU and CUHK. HKU's technology transfer office and its business development arm, Versitech, were formed in 1998 to capture commercial value from upstream research conducted at the university. Initially focused on the areas of information technology and engineering, Versitech turned to biotechnology during the early 2000s. According to Senior Manager Andrew Chan, over 80 percent of the patents managed at present by Versitech are in the life sciences field, even though most of its licensing and business activities continue to be in either IT or engineering.

The CUHK's Technology Licensing Office (TLO) was also established in 1998 and, similar to HKU's Versitech, the majority of disclosures and patenting activity in the CUHK TLO is in the area of biotechnology. The TLO has been very active in technology transfer. For instance, in the fiscal year 2006/2007, of the 37 disclosures (not necessarily in the life sciences) made to the TLO, 29 of them, or 78 percent, resulted in a filed US patent application. That same year, 19 licensing deals were finalized, equaling 66 percent of the number of patent applications filed. Licensing income earned in 2006/2007 by the TLO neared US$1.2 million, or roughly 2.6 percent of total research expenditure. To put that output into comparative perspective, Massachusetts Institute of Technology's (MIT) income to research expenditure ratio in 2006 was 3.6 percent, while Stanford's was 8.8 percent. Both MIT and Stanford are among the world leaders in transferring technology. The average income to expenditure ratio, calculated from a survey of 155 universities, was just 0.9 percent, considerably lower than at the CUHK's TLO. In fact, between 1992 and 2008, the CUHK received a total of 288 disclosures, of which 203, or 70 percent, resulted in patenting activity. Of those patented disclosures, 87, or 43 percent, were transferred to industry in either a licensing deal or another mode of technology transfer (AUTM Licensing Activity Survey, FY2006).

Clearly, Hong Kong's efforts to enter into the biotechnology sector have been significant. Government funding in Hong Kong, while minuscule when compared with the levels of public investment in other advanced countries, has been disproportionately large in the life sciences. Institution building has been a priority. Nurturing upstream basic science research capacity, especially within universities, has also been a priority, reflected in Hong Kong's competitive output. At the same time, efforts have been made to translate upstream knowledge into commercially viable outputs. The "pieces" of a biotech industry have begun to emerge in Hong Kong. Commercializing biotechnology, however, has lagged. The prospects of creating a commercially viable bioindustry in Hong Kong remain terribly uncertain.

3. The China pull

Under conditions of technological and economic uncertainty, it is difficult to determine where best to allocate scarce resources, a conundrum that is all the more pressing in tiny Hong Kong where resources are very scarce. Resources, be they public or private, have to be allocated *somewhere*. Niches have to be discovered *somewhere*. But where? Though the Hong Kong government eschews vehemently the notion that the state ought to (or even can) "pick winners," Hong Kong's proximity to and political–economic integration with China mean that allocative decisions essentially have been made for Hong Kong. China's pull is simply irresistible. And that strategy has been formulated, in part, by Hong Kong's economic policymakers. China offers "low hanging fruit" for would-be bio-entrepreneurs based in Hong Kong. Hong Kong can take advantage of its regional economy.

Hong Kong's economic integration with China, and the pull of Chinese economic development more generally, is inevitable given its close proximity to China and the official handover of Hong Kong back to the Mainland in 1997. The Closer Economic Partnership Agreement (CEPA), signed by the central Chinese government and the Hong Kong Special Administrative Region (HKSAR) in 2003 and implemented the following year, is hastening the opening up of economic activity, especially trade, across the China–Hong Kong border. The Greater Pearl River Delta (PRD) region is the basis of Hong Kong's economic future and its ambitions for industrial upgrading. With respect to the life sciences sector specifically, the HKSAR "is poised to play a significant role in the development of biotechnology industry in China" (Chang, 1999). During the early 2000s, about two-thirds of Hong Kong's pharmaceutical and health care-related exports went to China (Nature, 2001, 5). Moreover, Hong Kong, given its global reputation as a services and logistics hub in Asia, is positioned to play a "supporting role" for biotech development in the Mainland (Frost and Sullivan, 2002). Its advanced health care infrastructure and world-class universities make Hong Kong an ideal place to "bridge" global life sciences industries with China (Wong, 2006, 221–222). Indeed, due to Hong Kong's small local market and other scale-related bottlenecks (such as the small local pool of R&D talent), Hong Kong needs China as much as China needs Hong Kong if the former British colony is eventually to realize its ambitions in the knowledge economy.

3.1. Traditional Chinese medicine

In the near term, Hong Kong's nascent life sciences industry looks to gain a significant foothold in the traditional Chinese medicine (TCM) market, especially the huge Chinese domestic market. To the extent that Hong Kong enjoys any comparative advantage in the human health care industry sector, it is in TCM manufacturing, a point emphasized in the 1997 *Made By Hong Kong* study (Berger and Lester, 1997). Ever since the TCM sector was highlighted by the HKSAR government during the late 1990s to be a key priority, R&D resources, both public and private, have been allocated for the "modernization" of the TCM industry. While there is no standard definition of TCM modernization, most understand the process to be rooted in the principles of evidence-based research and the integration of TCMs into standard pharmacopeias. That is to say, most efforts in Hong Kong to modernize TCM involve the extraction and isolation (at the molecular level) of the active ingredient(s) in traditional medicinal preparations, followed by the rigorous testing of such extractions. Extraction and lab testing are intended to demonstrate, through evidence-based research design, TCM's efficacy and safety. The Hong Kong Jockey Club Institute of Chinese Medicine is funding or participating in several R&D projects aimed at modernizing TCM in precisely these ways, most notably an ongoing collaborative effort with CUHK and Baptist University of Hong Kong to develop a TCM product to treat irritable bowel syndrome. The ICM at CUHK is similarly running clinical trials for manufacturers of TCM products.

Institutes and universities in Hong Kong engaged in TCM R&D are increasingly working with industry, in large part because most TCM firms are without the R&D facilities or research talent to carry out their own research in-house. For instance, Vigconic, a TCM manufacturer and part of the Luk Industries Group, contracts clinical R&D projects to local universities to gather data on the efficacy and safety of its products. Eu Yan Sang Chinese Medicines invested HK$10 million in 2000 to seed a collaborative project with the CUHK, specifically a pharmacological study of the firm's "meno-ease" product. The initiative was co-funded by the government's ITF. For the CUHK, the research collaboration with Eu Yan Sang was about demonstrating "proof" of the efficacy of traditional herbal formulas for mitigating the effects of menopause. From the perspective of the firm, the "modernizing" effort was intended to demonstrate to the market the product's efficacy and safety in inducing the body to naturally develop estrogen (rather than require the ingestion of the hormone). These sorts of R&D collaborations have become increasingly common among the larger, more established TCM firms in Hong Kong.

Hong Kong's TCM industry is reasonably well developed. It is estimated that there are around 100 TCM firms in Hong Kong. For most companies, their core business rests in manufacturing, marketing, and sales. To be sure, the local market in Hong Kong is rather significant. A growing percentage of people in Hong Kong opt to consult a Chinese medicine practitioner before going to a Western physician.[2] The majority of people in Hong Kong are estimated to have consulted a TCM doctor at least once, it not regularly, for the treatment of common ailments and illnesses (Lau, 2000). However, the limited size of the local market in Hong Kong and the large number of local firms mean that the *future growth* of Hong Kong's TCM sector requires expansion into the Chinese market. And the fact that until recently TCM regulations were relatively lax, has meant that excessive market crowding among local firms, especially small manufacturers without Good Manufacturing Practices (GMP) certification, is very pronounced. Small firms may not survive in the long run, while large firms are constrained in the ability to expand their operations into the Chinese market.

PuraPharm is one Hong Kong TCM firm that has gained a sizable market share in China. Using proprietary technologies to derive novel molecular formulations from traditional herbal mixtures, PuraPharm has built a core business around the production of TCM "granules," or the de facto isolation and manufacture of key medicinal ingredients. In order to gain closer access to raw materials (herbs), PuraPharm early on established manufacturing facilities in Guanxi Province, China. As one of six firms to be granted a license from China's State Food and Drug Administration (SFDA) to produce TCM granules in China, PuraPharm's primary market has been the Chinese Mainland. China accounts for 65 percent of the firm's sales; Hong Kong, on the other hand, accounts for just 20 percent.

The case of PuraPharm is the exception, however. The reality is that for most TCM firms in Hong Kong, even the larger ones, the vast majority of sales (70 percent and higher) is in the local Hong Kong market. Though many firms are beginning to allocate resources to R&D and to modernizing TCM, most continue largely to focus on manufacturing, marketing, and distribution within Hong Kong. There are several reasons for this. First, most firms in Hong Kong are not GMP certified and are thus restricted in their ability to export products, even to China.[3] According to local TCM manufacturers, attaining GMP certification is very costly and most firms are without the resources to upgrade facilities. Second, the investment community in Hong Kong has shown little interest in the TCM sector, despite the government's attempts to highlight this potential growth industry. Most TCM firms are small family-based operations. Only a handful has benefited from angel investors and even fewer have been acquired by a larger industry group or holding company. Firms are without the resources to expand. Third, despite efforts to integrate and harmonize regulatory regimes in Hong Kong and China more closely, registration of TCM products in China, especially if they are not registered as a health food, is extremely arduous for local firms. Not only is the registration process in China expensive (reportedly up to five times the investment required for registering a TCM product in Hong Kong), it is also very time-consuming and sometimes arbitrary. Clinical trials can only be conducted at certain SFDA-sanctioned centers, and inside "connections" are often required to gain permission. And as the SFDA attempts to rebuild its regulatory reputation after several scandals were revealed in 2005 and 2006, the registration pipeline has slowed considerably, reportedly taking three times longer than before.

3.2. Clinical R&D

Nature noted in 2006 that in "the life sciences, Hong Kong provides a strong clinical research infrastructure" that can be utilized to bridge Hong Kong to emerging biotech industries in China. Hong Kong's hospitals and universities are world-class in terms of research. Its intellectual property regime is considered to be very strong, both in legislation and enforcement. Legal transparency is a high priority in Hong Kong (Nature, 2006, 221–222). In other words, Hong Kong can, in the medium term, position itself to be a principal site for conducting clinical R&D for both multinational and Chinese firms and labs.

During the mid 2000s, after several years of negotiations, the Chinese SFDA and the HKSAR announced that data generated from clinical trials that had been conducted at Hong Kong's university-based hospitals would be recognized by Chinese regulatory authorities, making Hong Kong the only location outside of the Mainland approved by the SFDA to permit clinical data transferability. Hong Kong could therefore capture a lucrative link on the biotech commercial value chain, especially as more and more multinational firms are looking to find a suitable gateway into China. Firms are assured by Hong Kong's enforcement of IP protection as well as its high-quality clinical research and data collection capacities.

Despite commercial promise in this specific niche, clinical R&D capacity in Hong Kong needs to be strengthened. Most clinical trials conducted in Hong Kong tend to be phase 3 trials, which are the least risky and capture the smallest value-added. There has been little effort to solicit earlier phase 1 or 2 trials, which would require considerably greater expertise in clinical research but which would also capture more economic value. Improvements in this regard need to be implemented relatively quickly, however, as Hong Kong's window to gain a market share vis-à-vis China in the clinical R&D business is closing fast. Industry observers note that labs in Beijing and Shanghai are rapidly developing their capacities to handle clinical trials for both domestic and international firms. Most important, the Hong Kong government needs to hasten regulatory harmonization with the Chinese SFDA regarding clinical research and data transferability. The 2005 decision by the SFDA took years to negotiate, and it remains unclear how long it will be before the agreements are actually implemented. The scope of the agreement is also quite narrow, limiting the range of foods and drugs that qualify under the agreement and also limiting the specific research institutions in Hong Kong that can carry out a clinical trial. Work remains to be done over the medium term – in both Hong Kong and China – if Hong Kong is to leverage its relationship with China and its infrastructure for realizing commercial gains from clinical R&D.

3.3. R&D collaboration

Over the longer term, there will be even more opportunities for Hong Kong-based biotech firms and research centers to collaborate with China's rapidly growing life sciences industry. There is considerable R&D talent in China willing to work in Hong Kong. During the late 1990s, roughly half of the life sciences researchers based in Hong Kong originated from China (Nature, 2001, 5). Institutional collaborations are being forged by Hong Kong and Mainland Chinese labs. The flow of talent and knowledge is not just one way, from China to Hong Kong, however. Hong Kong-based firms are looking to locate their R&D operations in China in the near term. For instance, Hai Kang Life Sciences (formerly Hong Kong DNA Chip), a Hong Kong-based start-up developing a novel lab-on-a-chip product, moved their principal R&D facilities and university collaborations to Beijing University early on in the company's development. In order to generate a short-term revenue stream, Hai Kang Life Sciences performs Genetically Modified Organism (GMO) testing in its Hong Kong facilities. Meanwhile, the firm's core technology business, the development of DNA chips, takes place in China. SinoMab is another commercial example of Hong Kong–China collaboration. SinoMab, a Hong Kong-based firm currently based in the Science Park, was initially founded on novel antibody research that had been conducted in China. The firm has since developed a re-engineering technology used on antibodies in order to identify and develop new drug candidates. While much of the firm's intensive R&D takes place in Hong Kong, SinoMab has established a GMP-certified pilot plant in Shenzhen. The firm also contracts out pre-clinical R&D to various research institutes in China.

Sustained R&D collaboration between researchers in Hong Kong and those in China is the basis for realizing Hong Kong's longer-term ambitions of becoming a biotechnology industry *innovator*. Hong Kong readily taps into China's significantly larger (and cheaper) talent pool. Hong Kong-based firms can gain easier access to the Chinese market, especially if clinical and pre-clinical research efforts involve collaboration among labs and firms in both places. To be sure, China is producing good science in the field of biotechnology. The Beijing Genomics Institute (BGI), the world's third largest genome sequencing lab and the first to sequence the Chinese human genome, recently moved its operations to Shenzhen. The local government there has supported the BGI initiative by providing resources, attracting returnees from abroad, and building up Guangdong Province's life sciences R&D capacities more generally. Hong Kong can leverage these sorts of developments over the long term.

The key point is that while during the mid 1990s, the future of Hong Kong's life sciences industry centered on its ability to move up the manufacturing value chain, and specifically in the TCM sector, more recent developments during the 2000s laid the foundation for considerably more lofty ambitions over the longer term, which are to turn Hong Kong into a biotech innovator.

4. Discovered in Hong Kong?

Efforts to increase R&D collaboration among Hong Kong researchers and others in China and elsewhere signal that biotechnology stakeholders are endeavoring to develop Hong Kong's nascent life sciences industry further up the technology chain to capture greater value-added returns. Hong Kong is looking to become a biotechnology industry innovator. To date, there has been very little output in this regard. This is not to say, however, that there have been no cases of success in Hong Kong. For instance, the Biotechnology Research Corporation (BRC), the commercialization arm of HKUST's Biotechnology Research Institute (BRI), was formed in 2003 with a HK$175 million investment from the Hong Kong Jockey Club. The BRC formed a joint venture, TA Therapeutics (TAT) Ltd, in 2005 with US-based biotech company, Geron. R&D collaboration between Geron and the BRI was initiated in 2000. BRI and Geron are developing new telomerase activator drugs aimed at restoring cells in damaged organ systems, which are now in pre-clinical development (Business Wire, 2005). The plan is for TAT to bring new drugs to phase 1 clinical trial, after which time identified compounds may be licensed-out for further clinical development.

TAT is a rare success case in Hong Kong. And the reality is that product development in TAT still remains far from the market. Biotechnology innovation at the cutting-edge is a very long and uncertain process. Even though Hong Kong has demonstrated considerable capacity in upstream research, especially given its size, translating discoveries from the university lab bench into a commercially viable product or service is an extremely high-risk endeavor. In this respect, it is still too early to determine whether Hong Kong will be able to become a biotech innovator. However, it is clear that there exist several "gaps" that need to be addressed at the outset (Branscomb and Auerswald, 2001). These will be addressed in turn below.

4.1. Midstream capacity

Technological innovation requires the translation of upstream research into a commercial product or service. In the field of biotechnology, there are literally thousands of potential "leads," though the reality is that most will either languish or fail to make it to the market. Identifying viable leads is central to the innovation process; it is also the basis of uncertainty. The Hong Kong Science and Technology Park (HKSTP) enlisted the prominent Danish–Swedish Medicon Valley cluster to prepare a report on the state of biotechnology innovation in Hong Kong during 2003 and 2004. The HKSTP has since re-enlisted the Medicon Valley group to conduct further research in Hong Kong and the Pearl River Delta. Its task is to inventorize Hong Kong's stock of knowledge in the life sciences and to identify promising projects with commercialization potential. Identifying leads is only one challenge, however; applying IP and translating knowledge into a commercial product is another. This process is prone to market failure. Midstream institutions are needed to correct this.

Taiwan's Industrial Technology Research Institute (ITRI), considered to be the birthplace of Taiwan's globalized semiconductor industry, performs this midstream function in Taiwan's efforts at technology innovation and development. During the post-war period and into the current era of biotechnology innovation, engineers and scientists at ITRI have "brought" technologies closer to industry. The ITRI bridges the public and the private sectors by anchoring or participating in premarket R&D consortia among firms. It further develops new technologies in-house and licenses them out to industry, or in many instances spins out a new firm. In addition to revenues earned from licensing deals and from contract R&D services, the ITRI is publicly funded by the Taiwan government's Ministry of Economic Affairs. Because the processes of technological innovation are prone to market failure, industry is unwilling or unable to perform this expensive and high-risk midstream function. This is especially the case with small firms and in economies where there is no critical mass of biotechnology enterprises. Industry thus tends to be risk-averse. ITRI, and other similarly publicly funded midstream R&D institutions in Taiwan, such as the Development Center for Biotechnology and the National Health Research

Institute, bear some of the risk and much of the costs of market failure. Such midstream institutions compel otherwise risk-averse firms to enter into the biotechnology sector. They cannot overcome biotechnology's uncertainties, but they can help offload from industry some of the risk.

Hong Kong lacks these sorts of midstream mechanisms to help commercialize otherwise strong upstream research in the life sciences. The Hong Kong Institute of Biotechnology (HKIB) was created to fulfill such a midstream role, though due to the lack of resources, talent, and promising projects, the HKIB was reorganized into an incubation center further downstream. It had failed to fill the midstream gap. Though the major universities have institutionalized technology transfer offices (TTOs) or licensing offices (TLOs), most have not been effective in bringing university research to industry, especially to local firms. These offices lack resources, where often the biotech division is a one-person operation. Thus, TTOs and TLOs tend to be passive with respect to prospecting IP from within the university. They also have little experience in managing life sciences IP (i.e., coupling IP) strategically, which reflects an overall lack of bioindustry experience among university IP managers. And university technology offices tend to focus on out-licensing IP rather than leveraging the university's IP to create new spin-off firms (HKSTP biotechnology initiative internal report, 2004). In the fiscal year 2006, for instance, the CUHK TLO, one of Hong Kong's most successful and active university-based technology offices, though it filed 29 US patents and negotiated 19 licensing deals, did not create a single start-up firm. TLOs and TTOs simply lack the resources to jumpstart new firm creation (AUTM Licensing Activity Survey, FY2006).

There is no institutional equivalent of the ITRI in Hong Kong's biotech sector. Publicly funded research institutions dedicated to filling the midstream "gap" in the life sciences do not exist. When asked, most industry insiders in Hong Kong repeat the mantra that the Hong Kong government, unlike the state in much of the rest of northeast Asia, does not "pick winners." Reducing risk and uncertainty to incentivize otherwise risk-averse firms to enter into the biotechnology sector is considered *excessive* government intervention. However, things have begun to change, albeit slowly. The establishment of the Science Park and the creation of the Hong Kong Applied Science and Technology Research Institute (ASTRI), created in 2000 to facilitate the transfer of technology to local industry, are steps toward backfilling the midstream R&D gap in Hong Kong's innovation regime. Efforts toward building up Hong Kong's commercialization capacity are pointing in the right direction. Even so, with the exception of the Jockey Club Institute of Chinese

Medicine, the life sciences industry, especially those subsectors that are unrelated to TCM, continues to be excluded from the ASTRI mandate. More can still be done.

4.2. Investment

The valley of death – the "challenges faced by would-be innovators seeking to make the transition from scientific breakthrough to market-ready prototype" – is a prominent reality in all technology innovation endeavors. The absence of institutional mechanisms promoting midstream R&D widens what Branscomb and Auerswald (2001, 11–12) refer to as the "research gap." Meanwhile, the "financial gap," or the absence of "investment funds [needed] to turn [an] idea into a market-ready prototype," is just as problematic for potential innovators. Risk and patient capital is needed. The US biotechnology sector was fuelled primarily by competitive government R&D grants and a thriving venture capital (VC) sector. American-based biotech start-ups and small enterprises were also supported by investments from (and acquisitions by) the USA's large pharmaceutical industry and firms. Enticing risk and patient capital is among the most significant challenges to growing a biotechnology industry. Regional competitors Taiwan, Korea, and Singapore have all experienced the challenges of narrowing the financial gap. Simply put, finding risk and patient capital is difficult to do in an inherently uncertain industry such as biotechnology, particularly in countries that have little "track record" in the life sciences.

The problem in Hong Kong is not just about finding a critical mass of risk and patient capital, however; the challenge is identifying *any* sources of investment capital that are interested in the biotechnology sector. Private sector investment in biotechnology R&D is next to none in Hong Kong. In addition, there is no pharmaceutical industry in Hong Kong to make up some of the financial gap. Some firms, such as CK Life Sciences, one of Hong Kong's most successful biotechnology firms, might benefit from an angel investor, who, in the case of CK Life Sciences, is Hong Kong billionaire Li Ka-Shing. Most firms, however, rely almost solely on personal out-of-pocket financing. Investable cash is in short supply in Hong Kong. Firms are thus constrained in how innovative they can be. The cash-burn rate in biotech is uniquely high.

Hong Kong's VC sector is uninterested and thus inactive in biotechnology. Though Hong Kong is home to one of the world's most vibrant VC industries, only one investment fund, the Morningside Group, has invested substantially in Hong Kong biotechnology. Most venture capitalists, if they are interested in biotech in the first place, tend to invest in firms abroad, notably in China. But even Morningside's investment "commitment" to biotechnology in Hong Kong is relative. Morningside's biotechnology portfolio lists 30 invested firms. However, just one is clearly based in Hong Kong, though there are four additional TCM firms with potential ties to Hong Kong. The remaining biotech firms in the Morningside portfolio are based in either North America or China, with 17 firms in the US.[4]

There are several reasons why venture capitalists have been slow to warm to Hong Kong biotech. First, the dearth of seed funding (both public and private) available for firms to demonstrate their commercial viability is a deterrent for venture capitalists who might otherwise be interested. As such, venture capitalists in Hong Kong tend to be risk averse when it comes to biotechnology. Second, they lack experience and the expertise to execute effectively the due diligence required to identify promising projects and firms. Investments also tend to be shorter term and thus less patient. The dot-com bust in the global IT sector of the early 2000s has altered investment strategies as venture capitalists shy away from high-risk, long-term endeavors. Hong Kong-based VC firms are waiting for a commercial "success case" in the local biotechnology sector (HKSTP biotechnology initiative, 2004). Third, government-funded VC, such as the ARF, often "leads" new investments and, in turn, induces follow-on investments from investors in the private sector. Government leadership of this kind in places such as Taiwan, Korea, and Singapore mitigates risk for follow-on investors. However, when the ARF ceased investing in new firms from 2004, private sector investors have been without the assurances of government leadership in indentifying firms and projects. As a result, investors have become even more risk averse.

And finally, there are few viable "exit mechanisms" for venture capitalists, or the means by which investors can realize returns on their investments. Most notably, the second board of the Hong Kong Exchange, the Growth Enterprise Market (GEM) board, is roundly criticized for its poor levels of capitalization, competition from the main board for listings, and the lack of turnover in trading. Consequently, the GEM board is not viewed as a particularly good exit mechanism for investors. In fact, its operation was suspended during the mid 2000s. Additionally, the absence of a critical mass of firms in Hong Kong's biotechnology sector means that there are few opportunities for mergers, acquisitions, and other types of strategic alliances, all of which are considered to be viable exit strategies for venture investors.

4.3. Critical mass

Innovation requires connectivity among actors. As discussed, the lack of midstream R&D mechanisms widens the gap among researchers and industry. There exists a financial gap in Hong Kong's biotechnology industry, as private sector investors are unwilling and unable to bring novel ideas downstream into industry due to financing constraints. Hong Kong's biotechnology sector also lacks a critical mass of firms, therefore constraining opportunities for inter-firm collaboration. Linkage among biotech firms – for instance, R&D collaboration and information sharing more generally, mergers and acquisitions, or the formation of strategic alliances – is critical to cutting-edge technological innovation. It is the basis of knowledge diffusion. Yet, as one bioindustry insider put it, "there are many biotech firms in Hong Kong, but there is no biotech *industry* in Hong Kong." Because of the reasons discussed above – the absence of midstream institutions and the lack of risk and patient capital – new business creation in the biotechnology sector has been slow in Hong Kong.

The lack of a biotechnology industry critical mass means that there are fewer firms able to receive technologies from further upstream. In other words, one of the main obstacles in the way of developing midstream R&D capacities in Hong Kong is the absence of firms that are capable of absorbing and assimilating new technologies. The technology licensing office at the CUHK, for example, tends to license-out its life sciences IP to foreign firms rather than to the few local firms engaged in biotechnology R&D. Hong Kong's small biotech industry also means that firms have fewer opportunities to gain bioindustry experience. The movement of people and talent among firms is constrained in Hong Kong, though as Steven Casper (2007) describes in the case of the San Diego biotech cluster, it is precisely learning gained from the interaction of clustered firms that contributed to the commercial success of biotechnology in the US.

Limitations in Hong Kong's biotechnology industry are not just rooted in the small number of firms, however, but also in the weak bases of connectivity among them. Weak or non-existent inter-firm linkages constrain information exchange and learning, as well as opportunities for partnering and R&D collaboration. According to most stakeholders, the Hong Kong Biotech Association has become very passive. Recent efforts by the HKSP have attempted to correct this. The HKSP hosted Hong Kong's inaugural Bio-Exchange partnering event in the spring of 2008, which attracted 180 attendees and 28 partnering initiatives; the significance, of course, is that the inaugural event took place just two years ago.

5. Conclusion

On balance, Hong Kong enjoys a sufficient though nascent base for growing a viable biotechnology industry over the long term. Basic and upstream research capacities in Hong Kong are quite strong in the life sciences. Hong Kong has been able to attract, and to a limited extent train, a high-quality corps of life sciences researchers. With respect to biotechnology commercialization, renewed efforts in Hong Kong's biotech communities suggest some great potential. Hong Kong can – and should – take advantage of "low hanging fruit," leveraging its relationship with the Mainland while also diversifying its existing industries (i.e., electronics, IT services) toward health technology applications. The biotechnology towers in the HKSP have quickly filled with both local and foreign-based tenants. The Hong Kong Biotechnology Association has also sought out a new leadership team in an effort to renew up- and downstream linkages. The HKSP has, in the past few years, hosted several investor forums for global and Chinese firms. The international community has begun to recognize these efforts. Hong Kong's nascent biotech sector has attracted interest from European investment consortia. In addition, biotech "stars" from the US have created subsidiary R&D operations in Hong Kong, both drawing on and developing further local R&D talent.

Still, the reality is that Hong Kong's biotechnology base is small-scale and, as such, Hong Kong will continue to face several challenges as it looks to grow its biotech sector further downstream. One distinct advantage for Hong Kong, however, is that the China market looms large. Given its close proximity to and cultural, linguistic, economic, and political "closeness" with the Mainland, Hong Kong can take advantage of its relationship with the PRD and China more generally. Hong Kong ought to look beyond the TCM subsector, and especially its past focus on manufacturing, if it intends to capture higher value-added gains in the life sciences industry. Focusing solely on TCM will preclude the growth of new businesses, as the local TCM sector in Hong Kong is already quite saturated. In fact, it could benefit from some consolidation among firms, led by those with upgraded GMP facilities.

The largest challenge facing Hong Kong's nascent life sciences sector is its weak capacities in midstream R&D and biotechnology commercialization. More specifically, Hong Kong faces serious "gaps" in translating upstream research to industry, in financing, and in inter-firm connectivity. The 2004 HKSTP biotechnology initiative internal report prepared by the Danish–Swedish Medicon Valley cluster stated that "if Hong Kong wishes to do so, it can now establish a biotech cluster that includes a biotech industry" (HKSTP, biotechnology initiative internal report, 2004, 5). The key point is the allusion to the political, economic, and commercial will in Hong Kong needed to grow a viable local biotechnology industry. It will not be automatic. Hong Kong's biotechnology base, as it is today, is insufficiently developed.

Gaps persist. To leverage its proximity to the huge Chinese market and the Asian regional economy, for instance, Hong Kong needs to harmonize with China its regulatory regimes governing clinical R&D, clinical data transferability, and product registration. Investment, particularly seed funding for early start-ups, needs to be encouraged in order to help struggling firms narrow the gap during the early commercialization process. And finally, an institutional equivalent of the ASTRI solely dedicated to biotechnology needs to be created and funded in part through public resources. This speaks to a more significant reorientation of the Hong Kong government more generally. The Hong Kong government, inspired by its laissez-faire model of post-war economic development, adamantly maintains that it does not "pick winners." Having the government help firms bring products to market, or provide resources for public labs to translate discoveries into applied technologies, is understood by policymakers to be unnecessarily interventionist. However, it needs to be emphasized that government intervention for the purposes of offsetting market failure in technological innovation is not the same as picking winners per se. Even "lean" states such as those in the US invest heavily in inherently uncertain sectors such as biotechnology. The US National Institutes of Health (NIH) allocates about US$30 billion each year for life sciences R&D. Decision-makers in Hong Kong must therefore make a cognitive leap to appreciate how public resources allocated for cutting-edge technological innovation and commercialization are not equivalent to intrusive industrial policy.

Notes

1. The Innovation and Technology Commission (ITC) defines "biotechnology-related companies" as comprising "mainly healthcare-related companies with business in pharmaceuticals, medicinal or healthcare products of traditional Chinese medicine origin, and medical devices and diagnostics." The ITC definition of a biotechnology company does not stipulate that the firm be engaged in R&D.
2. Interview with Abraham Chan, CEO, PuraPharm, Hong Kong, 12 June 2008.
3. One estimate indicates that fewer than ten Hong Kong-based TCM firms are GMP certified.

4. Data compiled from the Morningside Group website, www.morningside.com, accessed October 2008. According to officials at the HKSTP, Morningside has funded three companies inside the Park, though they are not listed currently on the Morningside website.

References

AUTM US Licensing Activity Survey FY2006, internal document, Chinese University of Hong Kong, Technology Licensing Office.

Baark, Erik, "Innovation Policy Forensics: An Analysis of Biotechnology in Hong Kong," paper presented at the DRUID conference, *Dynamics of Industry and Innovation: Organizations, Networks and Systems*, Copenhagen, 27–29 June 2005.

Berger, Suzanne and Richard Lester, eds, *Made by Hong Kong* (Oxford University Press, 1997).

Branscomb, Lewis and Philip Auerswald, *Taking Technical Risks: How Innovators, Executives and Investors Manage High-Tech Risks* (MIT Press, 2001).

Business Wire, "Geron and Hong Kong's Biotechnology Research Corporation Form Strategic Alliance for Telomerase Activation Therapies," *Business Wire*, 2 March 2005.

Casper, Steven, *Creating Silicon Valley in Europe: Public Policy Towards New Technology Industries* (Oxford University Press, 2007).

Chang, Albert Yen, "The Role of Hong Kong SAR in the Development of Biotechnology Industry in China," *Animal Cell Technology: Challenges for the 21st Century* (Springer, 1999).

Cyranoski, David Nature, "Putting the Pieces Together: Hong Kong," *Nature*, 26 July 2001.

Frost and Sullivan Asia Pacific "Hong Kong and Biotech, the Right Combination?" Pharmalicensing.com, 2002, accessed October 2008.

HKSTP, biotechnology initiative internal report, prepared by Medicon Valley for the Hong Kong Science and Technology Park, 2004.

Lau, Joseph, "The Choice Between Chinese Medicine and Western Medicine Practitioners by Hong Kong Adolescents," *American Journal of Chinese Medicine* (Winter, 2000).

Michael Hopkins, Paul Martin, Paul Nightingale, Alison Kraft and Surya Mahdi, "The Myth of the Biotech Revolution: An Assessment of Technological, Clinical, and Organisational Change," Research Policy (Vol. 36, 2007).

Pfeffer, Cary, "The Biotechnology Sector – Therapeutics," in Lawton Burns, ed., *The Business of Healthcare Innovation* (Cambridge University Press, 2005).

Pisano, Gary, *Science Business: The Promise, the Reality and the Future of Biotech* (Harvard Business School Press, 2006).

Tsang, J. C. and Y. L. Lo, "Biotechnology Development in Hong Kong: Infrastructural Support for Biotechnology and Related Industries," *Biochemical Education*, Vol. 26, pp. 150–152, 1998.

Wan, Frank, "Perspectives on the Biotechnology Industry in Hong Kong," *Asia-Pacific Biotechnology News*, Vol. 9, No. 18, 2005.

Wong, Grace, "Chinese Biotech: The Need for Innovation and Higher Standards," *Nature Biotechnology*, Vol. 24, No. 2, pp. 221–222, 2006.

10

Government Neglect and the Decline of Hong Kong's Integrated Circuit Design Industry

Douglas B. Fuller
Singapore Management University

1. Introduction: Taiwan's tortoise and Hong Kong's hare

Today Taiwan is recognized as the largest center of fabless integrated circuit (IC) design firms (firms which design but do not fabricate chips) after the United States, while Hong Kong is a bit player in this global industry. Fifteen years ago things looked quite different. At that time, Hong Kong was arguably ahead of Taiwan in terms of the technical sophistication of its integrated circuit industry. While Taiwan's large and cutting-edge pure-play foundries, firms which fabricate[1] but do not design chips, were already in existence and beginning to grow, Hong Kong at that stage had the technical edge in design activities. Furthermore, these design activities were eminently suitable for Hong Kong given its lack of a tradition of state support for large capital investments for industrial enterprises, just the type of support needed to jump-start IC fabrication plants (fabs for short) in emerging economies. Thus, 15 years ago one could easily have forecast that building on their respective strengths, Hong Kong would become a center of design to complement the fabs sprouting up in Taiwan, Singapore and Korea where generous state support for investment in fabs was forthcoming.

In the early 1990s, the pillar of Hong Kong's IC design activities was Motorola. Motorola had built up an impressive team of IC designers in Hong Kong to complement its manufacturing activities there. This team grew to such technical strength that it provided the lead on the Dragonball series of microprocessors in the 1990s. Taiwan had no technologically comparable design activities at the time of the design of Dragonball processors.

Beyond Motorola, Hong Kong also had Valence Technology Ltd, a very home-grown Hong Kong firm. Valence started out in 1985 and began by offering layout design services to NEC and Fujitsu. The firm quickly developed the capability to develop application-specific integrated circuits (ASICs) that captured product differentiation on a chip (Reif and Sodini, 1997). During the 1990s, the firm began to develop chips for Sony Playstation, and between 1998 and 2004 the firm made quite a lot of money designing chips for VTech and other system houses (firms that design and manufacture complete products, such as telephones) in Hong Kong (Interview with ex-manager of Valence). This firm was at least as technically sophisticated as the leading Taiwanese fabless firms of the 1990s, as demonstrated by its winning the Top 10 EDN Asia Component Design Award 2 years in a row (1995 and 1996). Indeed, the ability of Valence to serve major multinational corporations (MNCs) during the 1990s points to it having technical capabilities at least equivalent to those possessed by the major Taiwanese fabless firms of the time.

Despite the strong technical advantage Hong Kong enjoyed over Taiwan during the first half of the 1990s, Taiwanese fabless firms began to grow into relatively large-scale firms during the latter half of the 1990s. By 1997, Taiwan had four companies (VIA, SiS, ALI, and Utron) with over US$100 mln in sales. Significantly, three of these design houses (VIA, SiS, and ALI) were focused on PC chipsets, a very large market in Taiwan given the strength of Taiwan's PC manufacturing industry (ITIS, 1998, chapter 8, p. 11).

Despite Taiwan's advantage in having local PC manufacturers to consume its chips, Hong Kong was still a significant market for Taiwanese IC design firms in 1997. While 51.9 percent of Taiwanese design firm sales were in Taiwan, Hong Kong represented the second largest market with 29.5 percent of sales. Hong Kong actually represented a larger market share than Taiwan in consumer electronics ICs, with 55.1 percent of all Taiwanese consumer electronics chip sales (ITIS, 1998, chapter 8, p. 12). Thus, Hong Kong at the time still offered opportunities for fabless firms, especially in the consumer electronics space. Precisely because of Hong Kong's competencies in consumer electronics, Reif and Sodini urged Hong Kong to take advantage of the IC design opportunities afforded by Hong Kong's cluster of electronics systems houses (Reif and Sodini, 1997, 203–204, 208).

While the goal of this chapter is not simply to benchmark Hong Kong against Taiwan, this comparison does bring out questions worth asking about Hong Kong's IC industry. Simply put, what went wrong with

Hong Kong? Why did Hong Kong not build on its earlier strengths in the area of IC design? More importantly, how can Hong Kong recapture its past success and build a flourishing, globally competitive IC design? This chapter intends to answer these questions.

The chapter proceeds as follows. The first section presents an overview of the evolving structure of the global IC design industry over the last 30 years. The second section evaluates the current state of Hong Kong's IC industry and highlights its developmental bottlenecks. The third and final section will present the conclusion and some recommendations to remedy these bottlenecks.

2. The restructuring of the global semiconductor value chain

The global semiconductor value chain has witnessed a revolution in its structure over the last three decades.[2] In the late 1970s, the industry was still dominated by merchant IC firms (also referred to as Integrated Device Manufacturers or IDMs) that designed, produced and sold ICs, and vertically integrated electronics conglomerates that produced their own ICs which, more often than not, were used in their own electronics end-products (e.g., TVs, radios, and other electronics "boxes"). In either case, these firms controlled and conducted all the main activities of the IC value chain in-house, from designing the chips to assembling them into packages that could interface with other components and testing these final packaged chips.

Over the course of the 1980s, new firms experimenting with new organizational forms that tried to segment the vertically integrated value chain into discrete segments in order to concentrate on one of these segments emerged. For example, some of these newcomers aimed to design but not manufacture their own chips. However, these organizational forms were stymied by the fact that they were in a larger industrial environment structured to meet the needs of vertically integrated firms. Furthermore, even when the existing firms were willing to accommodate these firms by servicing them, there were organizational and technical barriers to sharing the necessary information to make outsourcing of most functions cost-effective.

During the first half of the 1990s, the technical barriers to vertical disintegration began to fall, as the advent of increasingly sophisticated electronics design automation (EDA) tools and other information technology innovations allowed for increasingly cheap and effective means of digitally transferring much of the necessary information needed to outsource various functions along the value chain. There emerged a co-evolution between these technical advances and the continued changes in industrial organization, as more and more firms began to focus on discrete functions or narrow sets of functions rather than pursuing vertical integration. As this new industrial structure matured, significant cost and time-to-market benefits accrued to firms embracing this focused approach, and these competitive advantages, in turn, pressured the remaining vertically integrated firms to shed functions.

The changes this co-evolution has wrought in the IC industry have been profound. Today, there are many firms, including large firms, that concentrate solely on design or fabrication or assembly and testing of ICs. Fabless firms that just design chips went from being just 3 percent of global market revenue in 1991 to 20 percent in 2006 (Hurtarte et al., 2007, 7). Similarly, pure-play foundries, firms which solely offer fabrication services for others, have grown from essentially zero in the late 1980s to 8 percent in 2006 (Hurtarte et al., 2007, 7, 26), and the revenues of foundries understate their importance in the global value chain since their share of manufacturing capacity is much larger than their share of revenues. Furthermore, the foundries and fabless firms have enjoyed a sustained growth advantage over the IDMs during the last two decades. Indeed, Taiwan's success in the IC industry is due to Taiwan being at the forefront of many of these organizational changes, particularly in creating spectacularly successful pure-play foundries.

The de-verticalization and segmentation of the global IC value chain opened up opportunities for smaller firms beyond those in Taiwan. Numerous successful fabless design firms have emerged in the US, Israel, and elsewhere, and pure-play foundries have sprung up across East and Southeast Asia. These successes suggest that the current global industrial structure would be amenable to new entrants in fabless design from Hong Kong, particularly with so much of the global IC value chain in close geographic proximity to Hong Kong.

A note of caution for the future is needed, however. Although the recent decades have offered tremendous growth opportunities, even for small start-ups, there are troubling trends that suggest the future may be a bit more difficult for smaller firms as the IC industry matures. Principally, the costs of design and fabrication are escalating faster than the market is growing. For example, at 45-nm process technology (the cutting edge process technology), the cost of process technology, plant and equipment for a 300 mm fab ranges from US$5 to US$6.4 b and design costs range from US$20 to US$50 mln. At 32-nm process technology, those costs are estimated to rise to US$13 b and US$75 mln respectively. These trends suggest that the industry will undergo further consolidation as IDMs, foundries, and fabless firms enter joint-ventures (JVs) to share costs (Hurtarte et al., 2007, chapter 14). Rising costs and consolidation may create larger barriers to entry than has been the norm over the last 20 years.

3. Findings: Current situation and developmental challenges

3.1. Interviews

This chapter focuses on IC design and ignores the two other main activities in the IC production chain, fabrication and the backend of assembly and testing. The reason this chapter concentrates on IC design is that Hong Kong has had historical competencies in this area and has not had any significant activities in the two other segments of the IC production chain in recent years. Furthermore, Hong Kong is unlikely to develop the capital-intensive fabrication segment where investments of at least several billion dollars are needed to build one current generation fab, and the backend of assembly and testing is relatively less technology-intensive so unlikely to provide a boost to Hong Kong's knowledge economy. Indeed, assembly and testing facilities have been concentrating in developing Asia over the last several decades as these activities migrate from higher-wage locations.

As the purpose of this chapter is to explore how to revive Hong Kong's IC industry, and promoting IC design appears to be the most promising answer, the research for this chapter targeted firms involved in IC design in Hong Kong. Of the more than 30 firms[3] involved in the IC industry in Hong Kong, 19 were interviewed. Among these 19, 17 are conducting IC design or are start-ups planning to do so shortly. Of the 19, 6 were MNCs and the rest were a mix of local firms ranging from large spin-offs from multinationals to tiny early-stage firms. Table 10.1 lists the MNCs and their activities in Hong Kong. Table 10.2 lists the Hong Kong firms and their activities inside and outside of Hong Kong.

Beyond the IC industry firms, the research also involved interviewing managers at key institutions within Hong Kong's innovation system, including the Applied Science and Technology Research Institute (ASTRI), the Hong Kong Science and Technology Park (HKSTP), universities, and various government agencies including the Innovation and Technology Commission (ITC).

Table 10.1 Interviewed IC industry MNCs in Hong Kong

Nationality	Hong Kong Activities	IC Designers	Analog/Mixed Signal (AMS) or Digital Design	Complete Design Flow	Design Metric	IC Centers (utilization)	ASTRI (utilization)
MNC	Design	2	Analog and digital	?	?	?	No
MNC	No IC design – only application design	N/A	N/A	N/A	N/A	N/A	N/A
MNC	Design	12	?	?	?	Use SPADE instead	No
MNC	Design	40 HK; 1–2 in China (SH)	analog	Complete except for new process tech	Advanced	Rarely	No
MNC	Managing backend of manufacturing	N/A	N/A	N/A	N/A	N/A	N/A
MNC	Design	14	Analog	Complete	?	ICDS	No

Notes: Question marks denote those questions firms declined to answer.

Table 10.2 Interviewed Hong Kong IC design firms

Nationality (Year founded)	Investment	Activities	IC Designers	AMP or Digital Design	Revenue	Complete Design Flow	Design Metric	IC Centers (utilization)	ASTRI (utilization)
HK (2003)	Self	Design	10 analog HK; 60–70 Mainland China digital and analog	Analog and digital	2007: US$6 mln 2008: US$60 mln	Yes	?	?	No
HK (1999)	Angel investor; IPO in 2004	Design	100–130 HK; 40 Singapore; 10 Mainland China	Mixed signal	2007: US$164 mln	Complete	?	Yes	No
HK (2007)	Self	Design service	1	Digital	No revenue yet	Mostly in-house	?	No chips yet so have not used	Talking with ASTRI
HK (2000)	Local investment company, Taiwan and German investors	Design	8	AMS	2007: HK$18.7 mln	Complete	0.35–2.0	Product Analysis; EDA Center, also SPADE	3 projects with ASTRI; 10 ASIC projects in 2007
HK (2001)	Started with several other ex-Moto guys	Design	10: 3TW, 7HK	Digital	2008H1: HK$25 mln	Complete	0.25–.22	ICDC, ICDSC; SPADE, testing also in TSMC	Considering purchase of ASTRI IP for SOC

HK (2006)	Still not completely spun-off from public institute	Design service	6	Mixed signal	?	Complete when needed	130 nm–65 nm	ICDC,ICDS; SPADE	Indirect – work for customer working with ASTRI
HK (2006)	Self	Design	2	Analog	2007: US$ 100k 2008: US$ 200–300k	?	?	?	No
HK (2005 as IC firm)	Self	Design in Shenzhen	4 designers in Shenzhen	AMS	HK$ 2–3 mln	Complete?	.6 bi-CMOS	ICDC	No
HK (2002)	Angel	Design	6	Analog, RF	2007: HK$ 3 mln	Can do complete	.6 bi-CMOS, 1.5 bipolar at BCD	No. Use SPADE and other outside sources	ASTRI design service and projects
HK (2004)	Angel – IC industry; internal revenue now	Design	4	RF and analog	2008H1:HK$3 mln		.18–1.0 CMOS	ICD, ICDS	No
HK (2005) (surveyed not interviewed)	?	Design	1	Analog	2007: 0	Complete	.35–.5	ICDS	?
HK	Planning stage	Planning stage	N/A	N/A	N/A	N/A	N/A	N/A	N/A
HK	Planning stage	Planning stage	N/A	N/A	N/A	N/A	N/A	N/A	N/A

Notes: Question marks denote those questions firms declined to answer.

3.1.1. Overall situation

Hong Kong has taken a large step backwards in its relative position in the global IC industry over the last 15 years. Freescale (formerly Motorola's semiconductor division) has drastically scaled back its design operations in Hong Kong and shifted most of this work to Suzhou, with ancillary operations in Shanghai and Beijing.[4] Motorola asked for state support in 2002 and reportedly Francis Ho, Secretary of the Commerce and Economic Development Bureau (CEDB), refused this request resulting in Motorola's decision to shift operations to Mainland China. Spin-offs from Motorola, both domestic (Solomon Systech) and multinational (On Semiconductor), have emerged, but these firms do not make up for the loss of this major design center. The MNCs that have moved in over the last decade claim that Motorola's ability to train good analog design engineers lured them to Hong Kong. However, none of these new operations are very large and analog design generally needs a few experienced hands rather than a large number of engineers so there are few opportunities for training large numbers of new engineers in these MNCs.

Reif and Sodini (1997, 211) called for the government to bring in MNCs to spur the development of IC design activities that Hong Kong desperately needed, but Hong Kong failed to pursue these activities, even as peer competitors in the region and beyond competed to take advantage of the globalization of R&D (Reddy, 2000). Even Taiwan, traditionally rather reluctant to offer MNCs attractive deals to locate in Taiwan, began promoting MNC R&D Centers during the past decade. Today, the MNCs have already established substantial offshore R&D sites for IC design in India, Singapore, and elsewhere. Unfortunately, Hong Kong has basically missed out on this opportunity.

The lack of MNCs has created fewer opportunities for engineering employment in Hong Kong and fewer opportunities for learning from large, technologically deep companies. Engineering graduates in Hong Kong have often gone abroad if they desired to pursue a career in engineering (see chapters by Wadhwa, Hart and Tian, this volume). If Hong Kong had been able to create a more vibrant labor market, in part through luring MNC design activities, engineering graduates could have stayed put in Hong Kong and created a virtuous cycle of a deepening labor pool, luring more firms to locate design activities in Hong Kong. Such a virtuous cycle possibly could even have spurred IC industry entrepreneurship as well.

Few domestic start-ups have emerged to follow in the path of Valence. This lack of technology entrepreneurship is primarily due to the lack of funding for such start-ups in Hong Kong in addition to the shallow talent pool. The other factor missing to spur more technology entrepreneurship is the link provided by returnees. Hong Kong has been unable to attract back the human capital in the form of returnees and investment capital from Silicon Valley and other foreign technology centers. Attracting human capital and investment capital from these global technology centers via ethnic networks have proven driving forces for technology development in Taiwan, Mainland China, India, and Israel (Saxenian, 2006). The dynamic combination of returnees and investment capital has not appeared even though a substantial number of engineers from Hong Kong are active in the technology sector in North America.

A comparison of the number of large global IC design houses from Taiwan, Mainland China, and Hong Kong demonstrates how far behind Hong Kong has fallen. Hong Kong has only one firm in the top 125 listed global IC firms, whereas Taiwan has 19 and Mainland China has 4 as shown in Table 10.3.

Hong Kong has tried to spur innovation through several new institutions created over the past decade, principally ASTRI and the HKSTP. While these institutions embrace the right ideas, they have not been able to overcome the bottlenecks in the form of lack of entrepreneurship and the lack of further development of MNC design activities.

These factors will be considered in greater detail below.

3.2. Funding

For the major centers of IC design across the globe, venture capital funding has been critically important to encourage firm creation.[5] The IC industry is fraught with risk, given its high technical demands and rapid product life cycles, and the main assets of many IC design firms are their human capital so banks are ill-suited to support new ventures in this area. Furthermore, the expenses to complete a chip design are increasing rapidly (see this chapter, pp. 245–246) so small amounts of seed funding no longer provide sufficient investment to cover development costs. In this context, ample angel funding and venture capital are required.

Unfortunately, in Hong Kong, the funding situation remains miserable. Venture capital (VC) firms simply are not interested in investing in technology enterprises in Hong Kong. Angel investors in the vibrant technology clusters of Silicon Valley and Taiwan often hail from within the technology sector itself. Hong Kong's lack of such a cluster creates a dearth of potential angel investors for the IC industry. The only

Table 10.3 Hong Kong, Taiwan and Mainland Chinese design houses among the global top 125 IC firms in 2006

Rank	Company	Revenue (USD)	Home	Rank	Company	Revenue (USD)	Home
40	Mediatek	1.6 b	Taiwan	97	Faraday	170 mln	Taiwan
54	Novatek	964 mln	Taiwan	101	Sitronix	138 mln	Taiwan
55	Himax	745 mln	Taiwan	102	ELAN	137 mln	Taiwan
58	VIA	658 mln	Taiwan	107	Richtek	132 mln	Taiwan
67	Phison	382 mln	Taiwan	109	Vimicro	128 mln	Mainland China
68	Realtek	381 mln	Taiwan	112	O2Micro	125 mln	Mainland China
72	Solomon Systech	322 mln	Hong Kong	113	Pixart	122 mln	Taiwan
73	Etron	322 mln	Taiwan	114	Holtek	119 mln	Taiwan
74	CoAsia	311 mln	Taiwan	116	Silan*	113 mln	Mainland China
85	SiS	243 mln	Taiwan	117	Silicon Motion	106 mln	Taiwan
95	Elite Memory	178 mln	Taiwan	120	Advanced Power	101 mln	Taiwan
96	Actions	170 mln	Mainland China	122	ALI	100 mln	Taiwan

Notes: *Silan has some IC fabrication capacity, but most of its revenue comes from IC design.
Source: Hurtarte et al. (2007).

other options are self-funding or getting customers to pay non-recurring engineering expenses (NRE) up front, but given the increasing expenses involved in designing complex chips, these options are also not very promising.

The funding experience of many of these companies is not very encouraging. Only three of the firms interviewed received any outside funding and this was all from angel investors. One holding company bought a small fabless firm with the idea of turning it into a captive design house for the firm's planned original equipment manufacturer (OEM) electronics expansion, but this type of tie-in between fabless design firms and other electronics firms has been rare. Since this acquisition just occurred 2007, it is too soon to hail it as a success. Valence itself has struggled for funding over the years. It received small amounts of funding from some of the Japanese firms that used its design services and then was given funding by Legend Holdings acting as an angel. In 1998 Valence was sold to SRS Labs, an American audio firm. One former Valence manager described the period under SRS Labs' ownership as the firm's golden age since the ties to SRS Labs provided more adequate funding. However, even SRS Labs was often reluctant to invest enough in IC development as it was not an IC firm itself. SRS finally sold the firm to Singapore-listed Willis-Array, an electronics supplier. Subsequently, the design team for this firm reportedly shrank to a small fraction of its former height of over 100 IC designers.

The two most successful domestic firms in Hong Kong are self-funded Appotech and Solomon Systech, which started as a local management-buyout/spin-off from Motorola. Appotech stands out as its founder, Chuck Cheng, returned from Silicon Valley after founding start-ups there. While returnee start-ups have been common in Mainland China and Taiwan, this type of start-up is rare in Hong Kong. Solomon Systech also received angel funding from the Solomon Group in Taiwan. However, Solomon Systech had a contract with Motorola to supply the firm with chips from one product line for several years after buying out that product line from Motorola so it was not the typical early-stage IC design firm without any revenue.

3.3. ITC/ASTRI

Most firms had little good to say about the ITC's Innovation and Technology Fund's (ITF) funding process. The ITC has too many layers of vetting committees of which only one necessarily has technical expertise in the relevant area. This vetting process takes too long for the research to keep up with product life cycles in the technology sector. The ITC vetting takes 3–4 months (according to firms that use ASTRI often) so this adds an intolerable lag in time-to-market for projects that should be completed in 6–9 months in order to get into the marketplace on time. However, some of ASTRI's own personnel suggested the whole time taken to be vetted by the ITC was more often 6–9 months (a whole product generation!) rather than 3–4 months. Indeed, there are five review panels.[6] The five panels are: internal, industrial review, technology review, ITC review, and board of directors. ASTRI personnel suggested that the process should just be delegated to the technology review board to speed up the process.

Problems are compounded because the ITC pretty much insists on the R&D Centers controlling the intellectual property (IP) they create by requiring that participating firms pay for 50 percent of the research in order to claim IP rights. This requirement creates an onerous burden for local firms in the context of Hong Kong's small technology sector. Some ASTRI personnel defended the ITC's IPR management pointing out that when the ITC intervenes in royalty negotiations (negotiations made at the start of projects), it is usually fair-minded enough to ask for a price that allows the projects to go forward. Another problem is that many small private firms in Hong Kong simply cannot readily absorb and commercialize the technology created in the 90:10 partnerships (research where private firms contribute 10 percent of the research cost). Given the low level of venture funding, even providing the 10 percent up front required by the ITC is a problem. Even for projects with a scale of HK$10 mln, smaller firms found it a burden to pay HK$1 mln up front before the project even starts, given that these projects can take 1 year, but currently if ASTRI is given money after the project is completed it cannot use that money for another project. The ITC has relaxed the 90:10 rule for ASTRI so private firms can donate less than 10 percent (see below).

According to ASTRI personnel, the ITC is somewhat more flexible than the private sector believes. For example, the ITC will sometimes provide start-up fees for ASTRI to work on a project while ASTRI finds a corporate sponsor to pay the 10 percent fee. This may be in line with the ITC's own report that not all funding is completely predicated on finding some industry funding. However, these "start-up" funds are quite small. In one case, the ITC gave the IC Group HK$350,000 out of a total project budget of HK$8 mln and a time limit of 6 months to find a corporate partner. The ITC also allows ASTRI to contract out its services with the customer paying 100 percent of the cost, but with ASTRI allowed to get an extra 10 percent from the ITC as long as the contract service work is related to projects being funded. In effect, through contract service work, ASTRI can receive 110 percent funding.[7] The ITC also lets ASTRI's industry partners make in-kind contributions that treat expenses the partners incur on the project as payments to the project in order to lower the cash contributions the industrial partners have to make. Furthermore, the ITC also allows ASTRI to get an average of 10 percent from a number of projects rather than a full 10 percent from each partner. However, there is a rumor that the ITC will up the amount industry partners have to pay from 10 to 12 percent.

Another major problem with ITF funding is that every grant over HK$21 mln needs approval by the Legislative Council (LegCo). This requirement is particularly burdensome for chip design because with this paltry sum one can barely cover the cost of getting a prototype. Neither Korea's Electronics and Telecommunications Research Institute (ETRI) nor Taiwan's Industrial Technology Research Institute (ITRI) have had to deal with such low levels of funding and lack of longer-term block grants. ASTRI's stopgap measure to deal with this problem is to ask for a review of all grants at once, even for those projects not starting right away. However, this method does not solve the problem of responsiveness, i.e., some projects need to be approved quickly so ASTRI can stay ahead of the curve in producing innovations.

The length of ASTRI IC projects is a problem from two perspectives. On the one hand, these projects are too long to keep up with changes in the marketplace. ASTRI projects typically last 1–1½ years so they need to be allowed to change midway through in response to shifting market demands, since market windows for products typically last 6–9 months. In this manner, the projects can be redirected to those market opportunities that will be available when the projects finish. However, the ITC does not tolerate any changes in the projects. On the other hand, these projects are too short to encourage ASTRI to do real research. The ITC's funding of 12–18-month-long projects does not encourage real forward-looking research so the ASTRI is mainly doing development work, but even with development work, the funding is too little to deploy technology into full-fledged commercial products. Of course, it appears many of the firms want development work from ASTRI more than they want research in any case. Nevertheless, if ASTRI wants to be like ITRI and serve to create and diffuse advanced technologies, it must make a more serious effort to do research.

ASTRI's IC Group consists of two different subgroups: one focused on analog and mixed signal design (called Portable Analog Mixed Signal Design or PAD) and a digital team called the Applied SoC (system-on-a-chip) Design team.[8] In 2007, the budget for PAD was HK$17 mln, with four IPs transferred to industry and six US patents filed. In 2008, the budget for the same group was HK$25 mln. The typical project has been 90 percent funded by ASTRI, and the Industrial Collaboration Projects (ICP) with a 50:50 split were only started in 2008. Even with the 50:50 split, ASTRI co-owns the IP but it cannot license it out. For PAD, the customer base in revenue was 100 percent Hong Kong in 2007 and a 75:25 split between Hong Kong and Mainland China through the first two quarters of 2008. PAD projected that its budget for 2009 would be HK$30 mln.

For the Applied SoC Design team, in 2007, the revenue stream was 70 percent Hong Kong and 30 percent Mainland. This team did not give any estimate for 2008 but suggested a shift toward the Mainland with the advent of 65 nm technology. The budget for this group was not disclosed.

PAD has 23 engineers. The Applied SoC Design team has 13 engineers. Both can do the complete design flow. To place these groups in international context, ITRI's SoC Technology Center has over 300 people, with the large majority being technical staff. In addition to being quite small, ASTRI's IC Group has suffered from high turnover according to one firm that has interacted with ASTRI.

Firms use ASTRI's design service since it functions essentially as a subsidized design service compared with the commercial firms. These firms pay 10 percent of the cost to use the design services. While helpful, this is a far cry from the major learning and diffusion role that ITRI has played in Taiwan. Still, ASTRI has transferred some IPs and is trying to create programs to help industry in other ways. One project is a mask set for a mixed signal SoC. Such a mask set normally costs US$1 mln, but as a number of customers (five to six thus far) want it, the firms only have to pay US$50,000–60,000 to ASTRI. One two-person team even spun off from ASTRI to become multinational Marvell's Hong Kong design team. The main problem with these small efforts is that they have been undertaken in a situation where more concerted, larger-scale efforts are probably needed to compensate for the other weaknesses of Hong Kong's quite small IC industry.

3.4. HKSTP/IC development center

HKSTP's IC Design Center and IC Development Support Center (these two centers will be referred to here as the IC Center or ICC for the remainder of this chapter, except when trying to distinguish a particular feature of one of the two since they both serve to support the IC industry with subsidized services)[9] offer a plethora of subsidized EDA tools, product analysis, and testing services. Local firms use these services quite frequently. Hong Kong University of Science and Technology (HKUST) Semiconductor Product Analysis and Design Enhancement (SPADE) Center's services are generally considered equivalent to or better than the services of the IC Development Support Center because SPADE has some better equipment, but the IC Development Support Center is cheaper. The MNCs only reported using the Park's Failure Analysis service and not the other services. Some MNCs use SPADE.

The ICC in the first half of 2008 received the rarely granted ISO27001 certification for its Information Security Management System. None of the IC centers in Mainland China have received this certification. Due to this certification, the ICC has been able to lure new services from two providers. IBM will serve the HKSTP and its firms with new process technologies (65 nm CMOS and 130 nm SiGe). The process technologies are on the US export control list and therefore not available in Mainland China. Previously, IBM never bothered to serve fabless firms with less than US$10 mln in revenue, but now plans to work through HKSTP to serve the Park's firms. The vast majority of the Park's fabless firms have less than US$10 mln in revenue. In a world first, Synopsys IP Trial has been made available to the IC center. It is claimed that this reduces cycle time from 18–24 to 6–8 months. With ISO27001, the ICC has virtual lock-in design rooms for the center's design tools that one can log in to from anywhere. However, Synopsys has set specific geographic boundaries from which one can log in to the system. Mentor and Cadence have no specific geographic boundaries for the use of their tools. ASTRI is using it and gets a 20 percent discount. Smaller firms get an additional 30 percent discount. This 50 percent discount represents what only the very largest firms receive worldwide.

The problem with these systems, in terms of boosting Hong Kong's IC industry, is that the clients appear to be Mainland firms with the exception of ASTRI (and even a fair number of ASTRI's clients are from the Mainland). For example, SWID from Chongqing is the sole user of the IBM processes thus far.

The ICC's subsidized services undoubtedly help the industry through providing the requisite industry infrastructure and lowering barriers to entry. Furthermore, the investment in these centers has been quite substantial by Hong Kong standards, with HK$230 mln spent from June 2003 through to November 2007. However, they are not enough to propel the industry forward given the other bottlenecks. Furthermore, the extensive use Mainland Chinese firms make of the services[10] begs the question of how exactly servicing Mainland Chinese firms benefits Hong Kong.

3.5. Returnees/expatriates

Despite the significant presence of Hong Kong engineers in Silicon Valley, very few have been lured back to Hong Kong to the technology business because the lack of an existing viable tech sector, lack of VC, and limited government support combine to offer few incentives to return. Dr Li, the founder of Kontel, came back originally to take care of the non-tech family business. The returnees in ASTRI (Ben Cheng and YK Li) either came back to ASTRI directly (Ben Cheng) or first went to try their luck in the Mainland IC industry (YK Li was at IP Core). The one significant entrepreneurial success story is Appotech's Chuck Cheng, who returned from the United States. However, despite having had a track record in founding fabless firms in Silicon Valley (e.g., Ubicom), Chuck had to self-fund Appotech and the bulk of his design team is in Mainland China because without VC investment it is too hard to hire a large design team in Hong Kong.

3.6. Cooperation with the Mainland

It is apparent that the Mainland firms are at least as active (and probably more so) than Hong Kong firms in utilizing the services of HKSTP's ICC. While this is not costly for Hong Kong because the Mainland central and local governments are subsidizing the Mainland firms to use HKSTP's services, this does not do much to develop Hong Kong's IC industry and is offering benefits to firms that are competing with or are potential competitors to Hong Kong firms. HKSTP has astutely taken advantage of Hong Kong's better IPR regime to gain the trust of a number of firms offering valuable services to IC design firms (e.g., IBM, Synopsys, Cadence, Mentor), but by providing Mainland firms with relatively equal access, one must ask if HKSTP is unwittingly undermining Hong Kong's competitive advantage by offering these services to firms that may compete with designers based in Hong Kong.

Although the ICC, under the auspices of HKSTP, is considered to be one of the nationally designated IC Centers (ICCs) of the PRC in the "seven plus one" formulation[11] in which ICC is the additional one alongside the seven national centers in Mainland China, what this means practically is that the ICC does not get any central government funding. However, it must also be said that most of the funding for the original seven national design centers comes from the local government – especially for the more successful design centers. It is also important to recognize that the national ICCs in the Mainland do not provide firms outside their jurisdictions access to their services except at higher, non-subsidized prices.

The lack of central government funding for Hong Kong is true across the ministries that deal with the IC industry and S&T matters in general. All of the central government officials approached about this subject said that promotional policies and funding from the central government for Hong Kong in these matters were essentially non-existent. What did exist, basically, was funding for Mainland Chinese firms to avail themselves of services in Hong Kong that were not available in the Mainland.

The Hong Kong-Shenzhen Innovation Circle Program and Hong Kong-Guangdong Technology Cooperation Fund provide funding for Mainland Chinese universities and companies from these jurisdictions to work with ASTRI to apply for ITC funding. However, personnel at ASTRI were concerned that the ITC would no longer accept even 50:50 funding for big Mainland companies, probably due to concerns about the need to spend the ITC's money to support local Hong Kong firms rather than large Mainland Chinese firms.

One MNC firm reported extensive cooperation with Zhejiang University, despite the fact that the MNC does not have any R&D in the Mainland. The firm described the decision to develop strong ties with Zhejiang University instead of with a Hong Kong-based university as having been made solely because "China was the flavor of the month" when the decision was made to look for a partner university in the region.

Hong Kong itself has dreams of jumping on China's technology bandwagon. The CEDB's *2008 Digital 21 Strategy* explicitly mentions cooperation with Mainland China on Chinese technology standards. Similarly, the ITC (2004, 27) envisions Hong Kong having an advantage in ICs through cooperation with the Mainland on its AVS and recently reconfigured WAPI standards.[12] The problem with this plan is that thus far Mainland China has failed to promote technology standards that have proved sustainable in the marketplace.

3.7. Labor supply

Most firms presented this issue as a classic chicken-and-egg problem. What seems to be true is that there are far more graduating electrical engineers than there are new hires in the IC business or related engineering fields. Many of the graduates either leave Hong Kong or leave the field. However, one major MNC expressed deep concern that HKUST shut down its MPhil in IC design, which trained quality IC designers, and replaced it with a much lower-quality, part-time MSc. This firm blamed the government for allocating too much funding for PhDs and not enough for MPhils. Of course, HKUST shut down the MPhil program because the labor market was so bad most of the graduates were leaving for work in the US and elsewhere. The firms concentrating on analog and mixed-signal design seem to have a decent supply because they do not need to grow very large teams and the legacy of Motorola left a pool of experienced analog/mixed-signal engineers in Hong Kong.

4. Conclusion

Hong Kong's IC industry woes are a telling tale indicative of the broader nested problems inhibiting Hong Kong's knowledge-intensive industries more generally. Given the previously strong capabilities Hong Kong possessed in IC design, the decline of this industry suggests that Hong Kong's cautious, positive non-interventionist tradition of policy-making is insufficient to maintain innovation activities let alone foster new ones. This is especially true when its newly industrialized economy (NIE) peer competitors have been very active in promoting IC design. Taiwan is the best known case, but Korea and Singapore have done well also. Singapore has even lured some MNC design operations away from Hong Kong. Korea has fostered a new group of fabless IC design companies, such as Mtekvision and Core Logic, to complement the formidable design capabilities of Samsung and Hynix.

Asia's emerging giants have also surpassed Hong Kong. India is now the preferred developing world destination for MNC chip design. As shown in Table 10.3, China has been able to link up with returnees to create innovative design companies as well (Saxenian, 2006).

Radical measures must be taken to encourage investment in the IC sector if it is to have any hope of flourishing in Hong Kong. To revive Hong Kong's IC industry, Hong Kong has to address the nested problems of poor entrepreneurial finance (see Chapter 6, this volume), a small pool of workers interested in the industry, and a weak innovation infrastructure. In order to do so, Hong Kong needs a big-push innovation policy similar to the big-push industrial policy discussed in the context of developing countries. Hong Kong also has to adopt some of the policies regarding returnees that have met with success elsewhere in terms of creating a larger, more dynamic workforce and industry structure.

Hong Kong's big push should provide much more generous funds for innovative activities that stimulate investment interest in IC design and other innovative activities, and lure more skilled workers to these fields of endeavor. One way would be matching funds for investment. Another would be to make public R&D funding more generous, flexible, and long-term oriented, as discussed in Chapter 5.

One measure should be to target the largest VC firms in the global technology sector (the top 20 or so) and/or the most active regional technology-oriented VC firms (e.g., Acer Capital, Walden) and offer matching funds for early-stage (seed and Series A) investment in Hong Kong's IC firms. For this to work, it is critical that the matching funds be predicated upon the investment decision by the international venture capitalists preceding the investment by the Hong Kong government. In other words, the investment by one of the targeted venture capitalists in a Hong Kong-based firm should automatically trigger investment by the Hong Kong government, but the Hong Kong government would never first choose in which local firms it would invest. In this manner, the Hong Kong government can ensure that the vetting process is done by the venture capitalists before using state funds. Obviously, the Hong Kong government would need to reach out to venture capitalists to explain this policy in order to increase their interest in the local market. A related measure could be targeted investment aimed at encouraging returnees to set up IC design operations in Hong Kong (see below). Unfortunately, given the current global financial crisis, this measure will take time to bear fruit as the venture capital market currently is dormant.

Encouraging venture capital would also help to solve the problem of which IC industry activities to encourage given the maturation of the sector and the likelihood of increasing barriers to entry. With the venture capitalists effectively exercising power of veto over government investment in this sector through this linked investment policy, the government would have a mechanism to prevent it from continuing to support an industry that venture capitalists recognize as offering too few opportunities for growth.

How would this policy of matching funds be different from the earlier, failed Applied Research Fund (ARF) scheme? First, the principle of automaticity needs to be in place to avoid the failures of ARF. In other words, whichever government organ is in charge of distributing the matching funds should first vet the venture capitalists and then automatically approve any investments the vetted VC firms make in Hong Kong as long as the investments are in the approved sectors (potentially other sectors should be promoted along with IC design). Second, the principle of speed of approval must be employed. The government organ in charge should be given no more than a week to veto the matching funds based on one of the two agreed upon investment requirements (location in Hong Kong and sectoral). If this 1-week deadline passes without a veto from the government, the matching investment should automatically be approved. Employing these two principles, the slow, bureaucratic approval process that hampered would be avoided. Third, a longer time horizon should be given to this VC matching fund. The government should not demand to see any positive returns for at least one decade because the point of this scheme is to promote industrial activity (i.e., success may show up in positive externalities not captured by the government's return on investment) and it often takes a long time for such early-stage investments to bear fruit. This third principle of a long time horizon will help insulate the matching funds program from unhelpful government interference demanding short-term profitability.

Hong Kong has underutilized terribly a great asset, namely the Hong Kong technologists living abroad. Returnees have played a significant role in the technology sectors of Taiwan, China, India, and Israel, and the fact that they are almost absent from Hong Kong, despite the obvious presence of Hong Kongers in global technology centers, such as Silicon Valley, needs to be addressed.

One way to do so would be to tie in certain venture capital matching funds mentioned above to luring experienced expatriate technologists to set up design operations in Hong Kong in return for venture capital. The matching funds might have to be made at rates attractive enough to lure expatriates home (i.e., the Hong Kong matching funds would have to demand less equity than the market rate), but would have to be made predicated upon outside VC investment as mentioned above.

Another important route to bind expatriate Hong Kong technologists to Hong Kong's technology sector would be for the state to set up, or at least financially support, a Monte Jade-like organization in Silicon Valley. Monte Jade Science and Technology Association, a Taiwanese–American organization based in Silicon Valley, played an important role in encouraging Taiwanese–American entrepreneurship linked to Taiwan (Saxenian, 2006). Hong Kong must seek a similar means to connect to its expatriate technology community in the US and use this vehicle to communicate about opportunities and government support, such as the matching VC funding, to lure expatriates to become more involved in Hong Kong's technology sector.

Another way to grow the labor market for IC designers and other knowledge workers would be to leverage Mainland China. While Hong Kong has made it relatively easy for educated Mainlanders to come to Hong Kong through its Admission Scheme for Mainland Talents and Professionals and the 2006 Quality Migrant Admission Scheme,[13] further targeted liberalization is required to bring engineering talent to Hong Kong. The Hong Kong government should make it much easier for Mainland Chinese and Taiwanese engineers to come to Hong Kong to work. A streamlined visa process for Mainland and Taiwanese engineers who wish to work in Hong Kong should be set up. For Mainland China, anyone with an MSc from an accredited microelectronics program should be allowed to come to Hong Kong under this streamlined process. For Taiwan, anyone with an undergraduate engineering degree from an accredited university should be allowed to take advantage of this streamlined process.[14] Following this, the government should sponsor an active recruitment drive at the major engineering universities of Mainland China and Taiwan.

The full-time MPhil program at HKUST should be reinstated. However, this reform should follow the other reforms for venture funding and returnees so that a growing demand for engineers is already in place when the training program restarts.

In addition to the calls for more flexible and longer-term funding for public R&D, Hong Kong has to confront the fact that it needs to leverage the assets it has vis-à-vis Mainland China. At present, Mainland Chinese cities do not offer free or subsidized services to Hong Kong firms that do not have operations in their cities. Hong Kong should treat Mainland Chinese companies in the same manner. For projects with Mainland firms, the funding should be recalibrated so that any non-contract service projects require that the Mainland partner place half of its engineers involved in the project in Hong Kong, with the exception that this requirement should not extend to those projects where the Mainland partner lacks any IC design capabilities. For example, Hisense's Shanghai-based chip design team is working closely with ASTRI. Since Hisense is a Qingdao-based firm, it could just as well have part of its IC design team in Hong Kong since the team is not co-located with headquarters in any case. With eased visa restrictions, Hisense could choose to bring its engineers to Hong Kong or recruit engineers in Hong Kong. This requirement would serve to bolster the sectoral cluster effects in Hong Kong as more engineers come to Hong Kong to work with ASTRI.

Similar to ASTRI, the IC Design and Development Centers should require that firms wanting to access the Park's services have the engineers who are using these services present in Hong Kong. HKSTP's superior services should serve as a lure for Mainland Chinese firms to set up design operations in Hong Kong.

The success of the major MNCs and the Taiwanese in the Chinese marketplace demonstrates that to sell into local Chinese manufacturers, chip designs must include complete turnkey solutions, that is, offer a complete reference design and software to accompany the chip. Local Chinese producers usually have weak design skills and only want to purchase chips from vendors who provide the complete reference design for them. In some areas, Hong Kong system firms have strong system-level design skills, but these skills need to be built up or else Hong Kong will never be able to compete with the Taiwanese, who offer very strong system design services for their customers. Funds need to be made available through ASTRI to target system-level design skills for those areas in the Chinese market where demand is high. Hong Kong's chip design houses could then access this system-level design service. ASTRI, private firms, and the other R&D Centers should jointly pursue opportunities to design-in chips in traditional manufacturing sectors where Hong Kong firms stand a chance of emerging as first-tier suppliers (see Chapter 9). In the medium term, the system-level design program should be phased out to push the fabless firms to hire the trained system designers themselves without subsidization.

Hong Kong had all the pieces to the puzzle to create a dynamic IC design industry: the skilled workers, the transnational technology community and electronics manufacturers. Yet, Hong Kong did not have sufficient public policy support to grow this sector. This benign neglect stemming from Hong Kong's positive non-interventionism has severely stunted the growth of what should have been a flourishing industry. Even in the country where the integrated circuit was invented, the US, the government, despite its laissez-faire tradition, gave tremendous indirect support through defense spending that created the flourishing American integrated circuit industry (Leslie, 2000). Virtually all the developing countries that have successfully entered the IC industry over the last several decades have had vigorous government support (Mathews and Cho, 2000). Hong Kong, by neglecting to provide such support, has inadvertently condemned a once promising sector to slow growth or even stagnation.

Notes

1. Fabrication is the front-end of the manufacturing process where circuitry is created on a wafer, usually made of silicon, to produce a semiconductor device.
2. This section is based on a number of works documenting the changes in the global semiconductor industry, including Berger (2005), Fuller et al. (2003), Fuller et al. (forthcoming), and Macher et al. (1999).
3. According to HKSTP, there are 38 companies (39 by HKUST's count because it counts ASTRI as a firm) in the IC industry within the Park. The list includes a number of firms that also have activities outside of the Park so it should be viewed as basically comprehensive for Hong Kong.
4. It is unclear whether the small design team in Beijing is in addition to the previous small design team in Tianjin or has replaced it. In any case, Freescale's website no longer reports a design team in Tianjin or Hong Kong although the website only reports "major" design centers, see http://www.freescale.com/webapp/sps/site/overview.jsp?nodeId=060A60.
5. Korea and Japan are exceptions to this rule, but they developed their industries through reliance on very large integrated device manufacturers (IDMs) combining design and fabrication. This capital-intensive model is not suitable for Hong Kong.
6. These five panels do not include the Legislative Council (LegCo) approval needed for projects over HK$21 mln.
7. There are conflicting accounts from ASTRI about whether this amounts to 100 percent or 110 percent. It appears to depend on how one does the accounting. In any case, ASTRI can leverage contract work paid by the private firm as that firm's 10 percent contribution to the ASTRI-led, ITC-funded project.
8. The ASTRI website still lists a third team called the Mobile Terminal and Multimedia team, but ASTRI personnel have confirmed that this team is no longer part of the IC Group.
9. The ITC (2004, 65), in discussing the role of the two centers, essentially lumped the two together by referring to them as the IC Design and Development Support Centre. The two centers are both managed, along with the Park's other labs, by the Business Development and Technology Support Division of HKSTP.
10. The Mainland Chinese firms usually obtain funding from regional and local governments in Mainland China to subsidize the cost of using the ICC's services.
11. There are rumors that Jinan will be added as the eighth national IC design base in Mainland China.
12. AVS stands for audio and video standard and is a codec (coder-decoder) compression standard for digital audio and video compression. WAPI stands for WLAN Authentication and Privacy Infrastructure and is a wireless local area network standard.
13. The former scheme grants entry to those who possess skills not readily available locally, and the latter has a points test through which potential immigrants compete to enter Hong Kong under a quota scheme (Office of the Chief Information Officer, 2007, 43).
14. The broader category for Taiwan is justified given the excellence of the engineering education available in Taiwan.

References

Berger, S. (2005) *How We Compete: What Companies around the World Are Doing to Make It in Today's Global Economy*, New York, Doubleday.

Fuller, D. B., Akinwande, A. I. & Sodini, C. G. (forthcoming) Digitization and the Rise of Greater China's IT Industry. In Fuller, D.B. and Rubinstein, M. (Eds) *Moving Knowledge: The People, Institutions and Firms behind the Flow of Technology from the United States to Greater China*, Manuscript.

Fuller, D. B., Akinwande, A. I. & Sodini, C. G. (2003) Leading, Following or Cooked Goose: Explaining Innovation Successes and Failures in Taiwan's Electronics Industry. *Industry and Innovation*, 10, 179–196.

Hurtarte, J. S., Wolsheimer, E. A. & Tafoya, L. M. (Eds) (2007) *Understanding Fabless IC Technology*, Amsterdam, Elsevier.

ITIS (1998) *1998 Bandaoti Gongye Nianjian* [*The 1998 Semiconductor Yearbook*], Taipei, Taiwan, Ministry of Economic Affairs.

Innovation and Technology Commission (2004) *Consultation Paper: New Strategy of Innovation and Technology Development*, Hong Kong, ITC.

Leslie, S. W. (2000) The Biggest 'Angel' of Them All: The Military and the Making of Silicon Valley. In Kenney, M. (Ed.) *Understanding Silicon Valley: The Anatomy of an Entrepreneurial Region*, Stanford, CA: Stanford University Press.

Macher, J., Mowery, D. C. & Hodges, D. A. (1999) Semiconductors. In Mowery, D. (Ed.) *US Industry in 2000: Studies in Competitive Performance*, Washington, DC, The National Academy Press.

Mathews, J. A. & Cho, D. S. (2000) *Tiger Technology: The Creation of a Semiconductor Industry in East Asia*, Cambridge, UK, Cambridge University Press.

Office of the Chief Information Officer (2007) *2008 Digital 21 Strategy*, Hong Kong, CEDB.

Reddy, P. (2000) *Globalization of Corporate R&D: Implications for Innovation Systems in Host Countries*, London, Routledge.

Reif, R. and Sodini, C. G. (1997) The Hong Kong Electronics Industry. In Berger, S. and Lester, R. (Eds) *Made by Hong Kong*, Hong Kong, Oxford University Press.

Saxenian, A. (2006) *The New Argonauts: Regional Advantage in a Global Economy*, Cambridge, Harvard University Press.

11

Hong Kong's New Creative Industries: The Example of the Video Games Sector

F. Ted Tschang
Singapore Management University

1. Introduction

Hong Kong has considerable creative and entrepreneurial resources, and the opportunity to build a vibrant set of creative industries, including in the new sectors such as video games, animation, and computer graphics. However, as it stands, some of these industries, especially that of the games industry, are fledgling in nature. Strong supporting institutions already exist, but it is essential to discover how industry can be better supported with existing and new resources – financial and otherwise. The opportunities are immense, but so is the competition. The new entertainment media sectors are growing at a faster pace than most economic sectors in many countries. At the same time, the expected global markets for new creative industries (especially games and animation) are considered to be huge.[1] India's NASSCOM estimated the global market for animation to be approaching US$50 billion, while one consultancy, DFC Intelligence, reported that the total global games market (including PC, online, and console) would rise from US$33 billion in 2007 to US$57 billion by 2009.[2] At the same time, investments in virtual worlds, including associated technologies and social networking sites, in the US alone have been in the few hundred million dollars range per quarter over 2007.[3]

1.1. Defining the new creative industries

From a policy and governmental perspective, the global interest in "creative industries" partly dates back to the UK's interest in defining and promoting this sector as part of a strategic growth and development initiative (1998).[4] The UK Department of Culture, Media and Sport (DCMS) defined 13 sectors as creative industries, but these included sectors as varied as software (including business software), antiques and collectibles, computer games, design, and architecture. The total contribution to the UK's GDP was US$112.5 billion or 5 percent of GDP (DCMS, 2001). In fact, the proportion that was represented by the newer sectors, such as video games, was far smaller than that for the larger sectors (video games being at US$1 billion, versus software at US$36.4 billion and design at US$26.7 billion), though the video games sector were presumably growing fast. Hong Kong's creative industries are said to involve 170,000 practitioners, 30,000 establishments and involve a total value added of over HK$53 billion a year, or 4 percent of the GDP.[5] While many countries have focused on promoting their creative industries, Singapore has targeted the new creative industries more strategically than most other countries. In Singapore, the creative industries themselves accounted for 1.9 percent of GDP (Singapore $2.98 billion); with the addition of the distribution industries, the percentage rises to 3.2 percent of GDP (Toh et al., 2003). From 1986 to 2000, Singapore's creative industries grew by an average of 17.2 percent per annum, compared with an average annual GDP growth of 10.5 percent. The Singapore government has committed funding of Singapore $500 million from 2006 for 5 years, to fund sectors such as games and animation in what is termed the interactive and digital media sector.[6]

It is worth noting that while some of the creative industry sectors are growing fast, they should be viewed not only as industries unto themselves but as possible competitive advantages or complementary industries to existing industries. For instance, computer graphic imagery or special effects are becoming well ensconced in major film budgets, while gaming is integrating with traditional toys and leveraging off the film industry's intellectual property. Finally, as Internet usage becomes more and more virtual world (VW)-based, VWs become targets for corporations interested in appealing to consumers by using VWs as new channels for marketing and interaction.[7] This interest in VWs is partly reflective of the IT sector's previous affair with digital media and e-commerce. While the dot-com boom of the late 1990s led to the crash in 2001, e-commerce has continued to grow, albeit as part of ordinary corporations' operations.

It is useful to paint certain characteristics of these selected creative industries first in broad brush terms – as shown in Table 11.1. The creative industries commonly referenced in Asian countries' policies include the video game, animation (largely three-dimensional),

Table 11.1 Characteristics of selected subsectors in Hong Kong's new creative industries

	Components	Form of user interaction	User objective	Purpose for business/ consumer
Video Games	**IT** (gaming technology), **content** (art and animation), interactive **design**	Active (interaction)	Challenge, socializing (MMORPG)	Product (may be service provider)
Animation	**IT** (computer graphic rendering), **content** (digital animation), **story scripts**	Passive (narration)	–	Product (may be service provider)
Computer Graphics	**IT** (computer graphic rendering), **content** (digital animation)	Passive (narration)	–	Service provider, own intellectual property develop-ment
Social media (incl. virtual worlds, social networking sites [SNSes])	IT (real-time rendering technology), **content** (VWs), **Web 2.0** (SNS)	Active (interaction)	Socializing (friends' networks)	Platform for users, creation (VWs); may complement other Internet sites or businesses
Design	Design aesthetics	Form (aesthetics) and (user) function	–	Service provider

computer graphics, social media, and design (including industrial design) sectors. This is especially the case in Japan, Korea, Hong Kong, and Singapore. It is this set that we refer to as the "new creative industries." It may be that video games and animation are particularly well known because of the popular culture and the press. Games can be further broken down to video or console games (typically played on a dedicated platform like the Sony Playstation or Microsoft Xbox) and computer games. (Hereafter, we refer to these collectively as video games or just "games".) Even though animation is not really as technologically driven as games, it seems to be holding a place as a valid industry for government policy to prospect for economic returns, perhaps because of its close tie in to popular culture and its appeal to younger generations. Social media is less well defined as a sector, but it is still as much about entertainment as it is about communication and social networking. Essentially, these sectors each encompass some of the aspects of being technologically enabled, creative, and "new" or "emergent" – or some combination of the three. They are also consumed in fundamentally different ways and involve different "components."

Animation: Animation has long and deep roots in both film and TV entertainment. However, animation itself has an older, two-dimensional tradition, as well as a newer three-dimensional one. While 3-D animation is known for its intensive use of computer processing, 2-D animation has also seen significant computerization over the years. There is a large global market for animation, but the outsourcing of animation – which many Asian countries also focus on as a growth path for their animation sectors – is restricted to the production stage (Tschang and Tsang, 2008). The alternative is to form one's own intellectual property, but with the exception of Japan's games and animation (i.e., the anime form) and Korea's online games industries, many other countries have found it difficult for their domestic products to cross the cultural barriers within other markets.

Games: The gaming sector is vibrant across the world, and online games have gained in popularity. Online games actually embody various other genres, with some online games being a mixture of casual (short-term play) and non-casual games (the former often being played "for free"). Sales data for the US indicate that console game sales went from US$7.4 billion in 2006 to US$9.5 billion (excluding online games) in 2007, of which console games accounted for US$6.46 billion and US$8.64 billion respectively. According to an Entertainment Software Association (2008) report, 22 percent of the most frequent US gamers pay to play online games. Of the total that play online games, only 11 percent play persistent multiplayer universe games of the sort like *World of Warcraft* (often labeled as massively multiplayer, online role-playing games (MMORPGs)). The total player population for *World of Warcraft* worldwide is about 12 million. Another estimate has it that total online game revenue in the US topped US$860 million in 2008, but this is still below the US$2 billion for the Chinese market.[8] This was driven by a user base of about 8 million households which visit MMORPGs monthly, of which about 3.5 million are "paying to play" in MMORPGs.[9] Another 9 million visit virtual worlds monthly, of which about 2 million are paying customers.

Computer Graphics: Another use of 3-D computer graphics related to 3-D animation is that of computer graphics imagery (CGI). This involves generating special effects for the film industry. Special effects have provided an increasing part of the value added to a typical US "Hollywood" blockbuster film. While MMORPGs are one of the original forms of virtual worlds, newer worlds are being created that are less game-like and more "social" (as illustrated in the next paragraph).

Social Media: Another manifestation of computer graphics is seen in "virtual worlds" which are platforms where social media and video game technology are coming together. The area of social media was originally represented by social networking sites (SNS), which were becoming the *de noveau* means of networking on the Internet. Facebook was best known for its embracement of Web 2.0 technologies which provided for a user-generated philosophy of development. However, virtual worlds like *Second Life* were taking this a step further by providing a new mode of 3-D avatar-based interaction and consumer-generated content. According to the Gartner Group, virtual worlds were set to become a pervasive feature on the Internet, and that as many as 80 percent of users could have a virtual presence by 2011.[10] At the same time, Gartner acknowledged that there was a high failure rate of corporate VW projects.[11] Whereas virtual world projects were very much in the forefront in Singapore – a large part of it derived from private sector funding – it appears that there is little corresponding activity in Hong Kong, other than SNSes like Zorpia.

1.1.1. The general structure of creative industries

There has been a growing academic literature on how the technologically enabled creative industries are organized (see Tschang (2007) for some sources), but the most relevant information for understanding the trends and structure of the industry is still contained within trade publications and "how to" books. Interestingly, many of the technologically enabled creative industries' key characteristics appear to mirror those of traditional cultural industries.

The critical features of creative industries (with economic and survivability implications for firms) are that the industries are hit-based, and that no one knows which product will be the next hit (Caves, 2000). This issue of predicting hits is even more complex with interactive entertainment, since consumers literally have to interact with the game or other product on a moment by moment basis (Tschang, 2005). The value chain for media is essentially segmented into the stages of development and distribution/sales, where development typically includes conceptualization, prototyping, (full) production (involving most of the human resources, especially the artistic), and final testing/debugging (in gaming projects) or post-production (in animation projects).

It is worth pointing out that many creative industries, and not only the traditional ones like writing and art, heavily involve intermediaries in the funding, marketing and distribution of the products. This is in part a segmentation of the value chain – one that allows publishers to be devoted to financing and distribution, and studios to be devoted to the creative or development work. The role that intermediaries serve in gatekeeping, seen in the arts (e.g., Caves, 2000), is present to an even greater extent in sectors like video games, since the complexity and costs of the latters' products require substantial amounts of funding. Intermediaries have significant influence over which products get funded. Unfortunately, intermediaries are not always as thoughtful or creative as the developers themselves, leading to tensions in the industry, as well as missed opportunities, or, conversely, potential opportunities.

While games, being information goods, have increasing economies of scale, they are also increasing in complexity, particularly with regard to content (e.g., art and animation) requirements. This increased complexity leads to an increased cost – both in producing games and 3-D animation. This occurs in part because consumers are increasingly expectant of higher quality in art and gameplay (i.e., the design for the type of interactivity) in games (Tschang, 2007).

Another feature particular to games is that they consist of multiple types of components. Innovation typically comes from innovation in the gameplay, but content (e.g., a new story "universe") with moderate amounts of gameplay innovation can also be used to create a "fresh user experience." This has been the experience in the Chinese gaming industry, where Chinese firms first licensed Korean technology, then imposed Chinese stories and story universes on their products to create a locally flavored product. This model works during the catching-up phase of industrial maturation, but innovations in gameplay have continued to come out of Korea, and most Chinese firms have continued to be imitative in their gameplay. Online games often degenerate into forms of role play or strategy play derived from PC games, or social play (which is more unique to MMORPGs) – the latter requiring less innovation.

A final point that should be made is that most game studios operate in a self-contained manner. Due to the iterative nature of the product's development, most studios create the game engine (or core programming code) themselves. Typically, the engine is intricately connected to the design and features of the game. While there are commercial engines available, if a studio is trying, at least partly, to be innovative, they still have to re-engineer the commercial engine's code considerably.

1.2. Hong Kong's new creative industry sectors

Hong Kong has also had an interest in promoting and growing the creative industries, in part because of the past industrial strengths the territory has cultivated in film, in design (including industrial design and fashion), and even in consumer goods industries such as toys. The film industry has enjoyed a long and storied past, though it had at one point become pigeonholed in Hong Kong-based action movies to some degree, and the lower levels of capitalization and "special effects" in past years had some observers suggesting that it had fallen far behind Hollywood or other emerging global centers of production (Teo, 1997).

Hong Kong has existing competencies and policy interests in at least four areas relating to the newer creative industries: video games, animation, computer graphics, and design. Generally speaking, these tend to be popular long-term and recent creative industries within Asia, particularly video games and animation in Japan, Korea, and China, and more recently, Singapore and Hong Kong, and animation in the Philippines (Tschang and Tsang, 2008). For the purposes of this chapter, we have limited ourselves to one narrow subset of this area – that of video games. Hong Kong's CGI, 3-D animation, and design sectors are more established, in part because of the territory's familiarity with how to take advantage of opportunities in acquiring and generating intellectual property (IP), and, in the case of CGI, the linkages with the film industry itself. To some degree, the production processes in these industries are well understood. It is when we come to the more technology-based industries such as games that the road to success is less clear, and the successes themselves, fewer. Yet, future possibilities – for example, that gaming and other new entertainment forms may fuse with "old" forms of entertainment or even converge with one another – suggest that it is prudent to continue investing in these new technologies and industries, and to build a stronger, diverse base across the creative industries (Table 11.2).

Our study is based on the secondary literature as well as a small but broad interview sample of about 12 interviewees (including

Table 11.2 Creative inputs and linkages with the new creative industries

	Link to popular culture	Line workers	Linkages with other industries
Video games	Culture provides content	Programming, design, art/ animation	Toys (e.g., a game representing a different way of playing the toy)
Animation	Culture provides content	Art/animation, writers	Leverage on IP from other media (e.g., comics)
CGI	Not linked to culture	Computer/ animation	Provide inputs to film industry, skill sets used in games, VWs
Social networking sites/virtual worlds	Usage varies by society	Programming, community management	Uses CG/animation skills, game design (to limited extent)
Design	May be culture-influenced or not	Designers	Strongly linked to products in other sectors

7 product-creating organizations). This is partly by design and partly a convenience sample (see Appendix for list of firms and institutes interviewed). The interview sample is overweighted in, and reflective of, at least the larger and most recognized firms in the video games sector – which are relatively small in number. I have also drawn comparisons with information gained from previous interviews of about ten online gaming companies in China. Focusing on the video games sector provides us with a lens into the problems of developing a particular sector.

1.2.1. Hong Kong government industry support initiatives

The Hong Kong government has traditionally provided a range of support for the creative industries, and more recently, the new creative sectors like games, animation and computer graphic imagery (CGI). Governmental initiatives can be classified by the dimensions of being educational, research-oriented, and industry-promoting. To support the creative industry sectors, Hong Kong developed various centers of competency, research and industry collaboration centers, and various university and vocational training programs.

The Design Center was established by the government in 2001 *"to establish Hong Kong as a centre of design excellence in Asia"* in part by promoting good design practices (HK Design Centre, http://www.hkdesigncentre.org/ accessed 9 June 2010). This largely involves promotional events, including exhibitions, as well as seminars and workshops. There are programs to connect students and emerging designers with established designers. Other initiatives to support the game industry have come from industry and institutions alike, including the Hong Kong Digital Entertainment Association.

Two of the tertiary institutions that support the industry are the Hong Kong Polytechnic University and the City University of Hong Kong ("City University"). Amongst other programs, City University has 2 and 3-year undergraduate programs in creative media and a Masters of Fine Arts program. The undergraduate intake is about 84 students per year. At the tertiary level, the Hong Kong Polytechnic University ("Poly University") has one of the strongest, if not the strongest, presence in the design and gaming sectors. Poly University has a range of undergraduate and graduate programs in various areas of design and multimedia. It also has a 2-year higher diploma in multimedia design and technology for which its intake is 136 students per year. It also established a Multimedia Innovation Center (MIC) in 1999. MIC's primary program is a Master of Science in Multimedia and Entertainment Technology. The 1-year degree connects students with both technology and design backgrounds. Students work on industry areas such as the following: Video Games and Online Entertainment, 2-D and 3-D Animation, Interactive Entertainment Systems, and Digital Video and Special Effects. Notably, this program's two core areas reflect to some degree the same emphasis as Carnegie Mellon University's (CMU's) renowned Entertainment Technology Center (ETC) Masters program, which brings arts and technology-trained students together. However, judging from CMU's experience, it may not be an easy proposition to operationalize such a program effectively.

Other Poly University research initiatives include game design and research (e.g., PlayLab, formerly known as MERECL or the Multimedia Entrepreneurial Research, Education and Creativity Laboratory), along with programs for "toy design," "public areas" design, and "Asian lifestyles" design. On the surface, each of these appears to be working at the cutting edge – our investigation of MERECL appears to bear this out. MERECL was jointly formed by the Innovation and Technology Fund (ITF) and the School of Design at the Poly University. It was formerly located at the Cyberport, but has since moved to the InnoCenter. There are about 80–95 people working there, and its employees are funded by the Poly University and the ITF, and cooperate with foreign companies. They work mainly on games or games-related technologies. The lab is for graduate students to undertake applied research and training, but they also hire graduates and industry people. The center focuses on commercial rather than industrial work, and has garnered praise from its North American partners.

In addition to Poly University and City University, two other institutions also carry programs in design, art and animation, and even video game design. These are the Institute of Vocational Education and the Hong Kong Art School. The Institute for Vocational Education's 4-year programs in games and animation carry about 260 students over the first 2 years. These programs mostly focus on basic skills for those not advancing to tertiary education. One issue may be the degree of professional training in the instruction. Generally speaking, educational programs will tend not to have qualified industry professionals teaching – it may, indeed, be a case of those who are working as professionals not having the time to teach and provide their experiences in significant ways to the classroom. These programs provide certificate and diploma courses for a wide range of skills. From interviews with firms, there appears generally to be no significant lack of talent in Hong Kong, although at least one firm noted that "well-qualified" people were hard to find. The opposing side of this is that there may not be enough firms with the capacity to fully employ and provide experience to creative talents either, making this "a chicken and egg" problem.

Another major experiment was the Cyberport's incubator – IncuTrain. Since December 2005, this has incubated 62 "digital entertainment" companies with the help of HK$30.8 million of the government's ITF funds.[12] A total of 62 out of 108 proposals have been accepted, which appears to be a higher rate than that for some other incubators. Most of these firms have less than five people each, and most of them were formed by people from "other (traditional entertainment) industries." The specialisms of the companies being incubated include major types of game genres, animation, edutainment, tools, and technology. Less than one third were gaming companies, but there were no cellphone game companies as the entry barriers were too low and too many firms had tried to enter the business, with bad consequences for most of them (this point was reinforced in other interviews). As with many incubators, the Cyberport has focused on providing training, office space, strategic advice (often pulling in the existing industry players as advisors), and the facilitation of business partnerships for its incubatees. The incubator provides a maximum of 2 years of rent-free space (under a government grant), which is fully subsidized. They provide support (in kind) to a level of about HK$300,000 per year per company. Manpower and training covers a significant proportion of the budget. The incubator "had graduates" within its first year of operation.

More recently, Cyberport has been focusing on console games, and has worked with Sony to try to improve support for the incubatees. The incubator negotiated with Sony to obtain "developer kits" that the incubatees could use to develop games. However, there is a limited supply of these kits, as they come with training and other support. For instance, Sony sends its own people to train on design, programming, and production processes. A similar program had existed for Microsoft's Xbox, and Hong Kong was second behind Taiwan in terms of Xbox incubation. In 2008, the incubator admitted its last batch – of less than 10 companies (under the original funding). A lot of the original firms have "graduated," but most have found additional investments to take their ideas forward, although they continue to remain small in size. The fate of the incubator or its successor is now under discussion in the Legislative Council (LegCo).

1.2.2. Prospecting Hong Kong's new creative industry sectors

In the sections that follow, we will assess first, the potential markets for Hong Kong's firms, then the industry's and firms' capabilities, and finally, specific supporting institutions (e.g., MERECL and the Cyberport incubator).

2. The market for and orientation of Hong Kong's new creative industries: The case of games

Hong Kong gaming and other new creative industry firms have a choice of entering various markets. As it stands, most firms start out and continue to service the domestic market. The IDC estimated that the online gaming market in Hong Kong was about US$30.6 million in 2006 (IDC, 2007). Furthermore, 10 software/gaming companies of 30 interviewed reported revenue of about US$20 million from online sales, while 15 of 20 CGI companies reported US$21.6 million in revenue, and 12 of 20 comic and animation companies reported US$3.9 million in revenue. In comparison, the Hong Kong GDP was about US$188.6 billion in 2006. Companies interviewed in the IDC study reported that they perceived Hong Kong to be their highest growth market for: animation (at 37.4 percent), digital effects (43.1 percent), and gaming (37.9 percent); these being double those of the next largest markets (typically either in SE Asia or China). However, this is discordant with one other observation: that Hong Kong's small population generally does not provide considerable revenue prospects. We can see this in how all the more mature gaming companies that we interviewed – F Game, G Game, and D Game Assets – were focused on overseas markets.

The prospects for the new creative industries are mixed.[13] We will provide a deeper analysis of game markets later in this section, but first we will provide a brief on the other new creative industries. CGI and design appear to have strong local markets, with both CGI and design being wedded to other local industries. The prospects for animation appear mixed, although there is always the long-run potential for generating new IP. The local market for virtual worlds and social networking sites (SNSes) appears to be minimal for now. Many Internet users continue to use the more traditional discussion forums for communicating practical information, and wireless may also be a stronger communication medium. Thus, it may seem that the Hong Kong market is too small to sustain an SNS – judging from the one company that we interviewed, which started out locally, but quickly moved to acquire a larger client base over Asia proper. SNSes are generally very much dependent on users' relationships with one another. Even then, at the small scale they were at, it was not clear whether they had sufficient expertise to bring their service to the next level.

In certain creative industries such as games, markets have a rapidly changing nature. Many companies that used to work on cellphone games (including one that was interviewed) have abandoned that market as the barriers to entry were too low (and revenue proposition increasingly unclear). It would appear that many of the firms (both large and small) have gravitated toward online games, as this is the only way to capitalize on the Chinese online games market. There are no Chinese firms working on console games, and some Hong Kong industry participants (e.g., the incubator and its incubatees) see potential in these products, but the financial and human resources to take these firms to the level that the advanced console technology has provided for may not be available (see Section 4).

2.1. Some dimensions of markets for new creative industries

One critical issue faced in selling products across regional market boundaries is the market-specificity of the products. That is, unlike electronics and other high-tech products which are primarily based on functional uses and specifications, products such as games may be highly culturally situated, and may appeal more to specific cultures, including the culture of the designers. For instance, despite the great success of individual Japanese designers and games in the US market, most Japanese games are unable to appeal to, let alone sell to, the US. This raises the hypothesis that products may best be catered to particular markets by developers steeped in the culture of those markets. While games would appear to be an unusual form of media in that they are *interactive*, they are as dependent on how players "view" the content as they are on the players' interaction with the games' design. The strong influence of culture comes in the sorts of gameplay that a culture is comfortable with, in addition to the content (consisting of the "worlds" and "narratives") that the gameplay helps enact or is embedded within.

The markets for both animation and games can be divided up in different ways. For instance, the dimension of "refinement" is often used as an indicator of whether a video game is of a "triple A" (or "AAA") quality, where quality refers to the production values of the game. Similarly for animation, a Pixar movie has greater refinement, in terms of the production quality (the degree of fineness of the image), the technological enhancements to the animation "effects," and, perhaps most importantly, the degree of refinement to the story. It is not at all easy for animation to achieve a regional, let alone global, market, judging from regional experiences. For instance, the Philippines' animation industry – one of the largest animation outsourcing providers in the world at one point – has also produced limited domestic content that is mainly appreciated by audiences in the Philippines. More recent regional animation productions that have been marginal successes thus far include the Singapore-made animated feature *Sing to the Dawn*. On the other hand, the Hong Kong film industry has had considerable success in exporting to regional Asian and, to a limited extent, global markets. Similarly, the greater China and other East Asian film industries are enjoying increasing success in regional markets, at least for individual films and directors. A related problem also confronts game developers. G Game's head noted that it's not easy to cross the threshold from being a "local" game or unique game with content specific to a certain market segment, to one that is high-selling.

There are of course various game genres, and also various degrees of innovativeness in games (depending on the component in question). Innovation in content may not involve innovation in gameplay (e.g., games with movie-like qualities), while innovation in gameplay may require less in the way of content (e.g., casual games require less content and their developers can take higher risks in inventing new gameplay).

2.2. Hong Kong's (small) domestic market for games

The Hong Kong market for games exhibits one possible distinction from China's: unlike in China, console games are a significant market in Hong Kong, and some PC games also continue to be made and sold. Of 30 Hong Kong companies surveyed in 2006, online game developers represented the primary business of 30.9 percent, mobile games 25.5 percent, PC game developers 12.6 percent, and console game developers 4.4 percent (IDC, 2007). Interviewees from our study also indicated that while in the past, a number of firms focused on mobile phone games, many of these had exited due to the low barriers to entry, smaller market, and more intense price competition. Most of the remaining Hong Kong developers now appear to be focused on online games, paralleling a worldwide trend that started several years ago, and that has been catching on in China. The experience of one interviewed firm – M Game – mirrored this trend. They started out making mobile phone games, but as the market became crowded, they moved into online games.

Unlike for many producers working on technological or consumer products, the immaturity of the new creative industry products means that it is more difficult to manage an enterprise for export, and in fact the best hope for developers (i.e., development studios) to achieve consumer satisfaction may be to aim for domestic markets. However, larger markets such as Greater China can be very attractive, and may even be necessary to broach in order for firms to succeed. Specific firms such as Enlight Software in Hong Kong (which is very successful at making PC strategy simulation games) and Object Software in Beijing have managed to export computer games and online games respectively with reasonable success over the years. Enlight has done this largely through regular contacts with overseas customers and markets.

Gaming firms that we interviewed noted the difficulty of achieving market success through a purely domestic strategy. However, as with animation, the challenge of achieving success in regional markets was also highlighted in a few interviews, with interviewees commenting in effect that "HK culture doesn't sell well." This may be due to the localness of certain aspects of cultural content, but the notion of culture in games needs to be unpacked more systematically and carefully in order to appreciate why games sell or do not sell in other markets (see next subsection). It is also worth noting that while the challenge of developing innovative games is often considered to be caused by the need for production values, truly innovative games are based more on original gameplay (i.e., game design) than on content.

2.3. Beyond the Hong Kong market: China and other markets

2.3.1. The China and regional market for games

Despite the suggestion by firms in the IDC study that the Hong Kong market had potential, two of our interviewees felt that Hong Kong's games market was not growing. Taiwan's market is in a similar condition, as is Japan's. In the past, a number of Taiwanese games were exported to China as well, but not anymore. In general, it seemed easier to import Korean games to these smaller markets than to develop them internally. One interviewee observed that, with few exceptions, the bulk of the Hong Kong industry has been acting in the same manner as that in Taiwan, with short-term profit-minded thinking being common. Part of the problem is that previously local companies could never tell how much revenue a good game might make. Now, this is less of a problem because G Game and F Game have "demonstrated" that very good artwork and reputation can create success, and this has helped set benchmarks for others to aim for in the local and regional markets.

Like the Hong Kong market, or in fact any market, the Chinese market has its own peculiarities. Two interviewees observed that Chinese games do not have to be that good (quality-wise) to make it in the large Chinese market, but this is changing all the time, as Chinese firms are continually upgrading their technological capabilities. At the same time, the Chinese PC games "market" has all but disappeared. Large-scale piracy, approaching 90 percent to 95 percent or more by some accounts, has driven out most domestic (Chinese) PC games developers, and console games are all but non-existent. Mobile games continue to prosper in China, in part because of the larger market, but the biggest single trend has been the growth of the online games sector. This sector now earns US$2 billion annually and is increasing.

As with Asian entertainment in general, there would appear to be a regional market for digital entertainment like online games. The fact that Chinese games developers like Kingsoft appear to be having some success in countries like Vietnam and Malaysia, along with the earlier success of Korean games in China, suggests this is the case. In this way, one of the chief markets that many Hong Kong games developers have aspired to enter is that of China, but this market has not been easy to breach for a variety of reasons. One common observation is that many gaming companies produce for local markets, and these products might not be successful in China. Many interviewees believe that local (Hong Kong) tastes do not run the same way as Chinese tastes. One interviewee brought up the example of a local company producing a game based on local celebrities whose names would not be recognized as much in China. Another interviewee noted that Hong Kong had taken a long time to create a successful movie industry and to get into the Chinese market, and that it will be even more difficult with regard to games. The development cycle is 1.5 years, budgets will run into the HK$10 million range, and tastes may change in the meantime. This is high risk, and it is not clear if venture capitalists could put up with this, let alone figure out the returns. As such, there are no venture capitalists funding games in Hong Kong.

Another challenge to entering the Chinese market is to find distribution channels, and handling the cultural differences in game content, all of which means partnering with Chinese companies. There are also many state regulations, and the Chinese government's rules have changed a lot. Many interviewees noted the difficulty of dealing with the Chinese business environment (regulatory, consumer, and business competition), but G Game has apparently found a way to work with this, by working with the regulatory agencies and by setting up a development shop in China to "attune" their products to Chinese tastes.

2.3.2. Bringing Asia to the West

Some interviewees also noted that the "Western" exposure of Hong Kong developers means that they can also bring Asian cultural influences into Western games. Local designs have also tended to be influenced by the West. Having said that, one interviewee stated that he had often seen Hong Kong "things" in a design that did not sell well, even in the West.

The issue of whether products can cross-sell in other markets may ultimately come down to a fine tuning of: (a) what aspects of culture are "bought" by consumers; (b) the tapping into the cultural traditions of other larger countries (e.g., Japan and China), and (c) the combining of this with Hong Kong's own cultural tradition (that may already be expected by other countries' consumers). For instance, firms like the medium-sized F Game are attempting to combine Japanese-influenced anime with their own traditions – a highly unusual mix that takes great skill and experience. On the other hand, another company noted that their use of Japanese intellectual property to design an online gaming/website environment did not do as well in Japan as it did in other countries. This may require some combination of knowledge and business acumen in dealing with Japanese consumers and publishers.

2.3.3. The potential for outsourcing as a strategy

The services potential of certain creative industry products may also be expressed in outsourcing or exports. The ability to provide this capability depends on the degree of maturity of the industries and their firms. Animation has long been mature in the sense that the production stage can be wholly outsourced. For instance, first Japanese, then Korean and Philippine, animation studios have been undertaking the production stage of animation for American (and other) TV and film studios. The likelihood of this being a strategy for a small labor market, high-cost country like Hong Kong is questionable. The profitability of this as an industry-wide phenomenon is also questionable as it is difficult to sustain a large enterprise for a long period of time. Due to increasing costs and turmoil in the global animation marketplace, the Philippine animation industry has failed once before, losing a substantial number of its firms, jobs, and business. It is only now picking up again.

There was some question about the potential for outsourcing in games. The D Game case confirms what one interviewee (from MERECL) observed – that outsourcing can make sense as a strategy. However, in order to become outsourcing providers, enterprises need to be placed in strong partnerships with clients (much as D Game works with a Japanese company), need to know exactly what is required (i.e., have experienced the development), and also be able to handle the complex tasks of integration (of the different components).

3. Assessing the strengths and weaknesses of industries and firms

Having examined the market potential for each sector, as well as the idiosyncrasies of cross-selling in different markets, we will now turn to the capabilities of the industries. We will first focus on the gaming firms in subsections 3.1 through to 3.3, and then return to a brief comparison with the CGI sector in subsection 3.4.

3.1. Assessing size

The new creative industries sector in Hong Kong is very small, no matter how it is looked at. The sector was estimated, in various interviews, to be made up of several key game development studios, animation studios, and CG (special effects) "houses." There are a number of smaller enterprises in each sector (e.g., dozens of start-ups and smaller game developers, such as those developing cellphone games), as well as a number of intermediaries in each sector (e.g., games publishers). The IDC (2007) study commissioned by the China Game Publishers Association (Hong Kong), reported 260 digital entertainment companies (employing 4600 people), of which 30 percent were involved in computer animation and comics, 45 percent in entertainment software and gaming, and 25 percent in digital effects. The study noted that 70 of these companies had 1260 employees in total.[14]

Despite the published numbers, our interviewees estimated that there were no more than a handful of good games studios in Hong Kong, most of these being in the 10–20 employee range in size. There are, however, dozens of start-ups if the incubators and other smaller firms are included. There were as many as 40 start-ups from Cyberport's initiative, of which some were focused on animation and games.

We later assess the industry's capability with different means. Capability is a stronger measure of firms' maturity and strength in technology industries, including those such as software (e.g., Arora and Gambardella, 2005). In video games, while studios have remained small, the notion of a capable organization rests on the abilities of the creative team, involving everything from the top design leadership to the line employees at positions such as programming, art and animation, and level design.

3.1.1. The strengths of other countries' games industries

The strengths within the games industries differ from country to country. Japan's long-standing manga and anime artistic tradition has benefited its gaming companies through the artistic talent side (Aoyama and Izushi, 2003). Similarily, M Game's head noted that the graphics in Korean games were excellent, and that South Korea had a lot of money for funding companies, even in the early stage. In general, many interviewees felt that countries like Korea and Singapore provided strategic benefits to their firms. At the same time, many mature industries like those of Japan and Korea have the advantages of stronger firms, local publishers, and finance dedicated to the games and other creative industries.

3.1.2. A comparison with China

None of the firms that we interviewed thought, or could afford to think, on as large a scale as their Chinese counterparts. Perhaps one reason behind this was the firms' low levels of capitalization. This may be due to Hong Kong being a higher-cost country, or having either a comparatively more risk-averse financial sector or lower degree of financial resources than China. The problem is, scale matters. Beijing's

Perfect World (or WanMei), for instance, became a 600-person developer/operator in a few short years of operation. It has been estimated by one industry participant that to build a game capable of competing with *World of Warcraft*, a company would have to invest between US$500 million and US$1 billion (it is likely that a large proportion of this amount would be dedicated to marketing and operations). Online games require companies (and often the intermediaries) to be product developers, marketers, and operators.

In contrast to Hong Kong, China's online game development capability is considerable by almost any measure. As shown in Figure 11.1, Beijing has the largest number of teams of the seven major game-producing regions: 41 in 2007, versus 22 in Shanghai, 20 in Guangdong, 13 in Jiangsze, 12 in Fujian, 12 in Chuan Yu, and 7 in others. This is not the complete population, as the number of games studios is in the hundreds, and often includes many small studios looking to make a quick return on smaller products. These smaller firms are often imitative in nature and make poorer quality products (in part due to their weaker state or lower capability). Looking at various team sizes of the 41 Beijing teams, we find the median team size to be in the 40–100 employee band, and the largest number of teams to be in that same band. In addition, there are 10 teams in the 100–300 and above employee band. It is likely that these teams include staff dedicated to sales or to operations of the games, since many developers also "operate" the games for themselves. In contrast, China's PC games industry has been whittled down

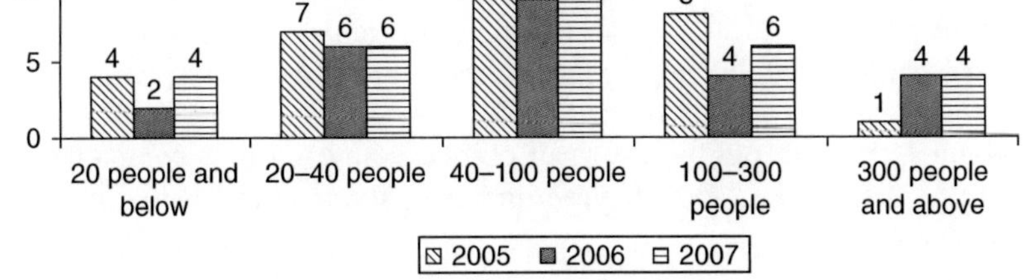

Figure 11.1 Size of teams in Beijing in 2005–2007
Source: www.17173.com, 2007.

by piracy to being almost non-existent, with very few, typically smaller, studios attempting to make PC games. The piracy afflicting these studios and imported games alike is still rampant however.

The increasing size of China's games industry is also reflected in the total number of teams. The number of development teams in Beijing grew from 27 in 2004 to 41 in 2007, while the number nationally grew from 73 in 2004 to 123 in 2006. Nationally, the number of professionals (including non-developers) in the industry grew from 5032 in 2004 to 21,034 in 2007 (in 2007, the proportion of developers to non-developers was 11,905 and 9129 respectively).

The largest Chinese firms are both publicly traded and privately owned. It appears that both types of firm have sufficient capitalization to underpin their high growth rates. One way forward for Hong Kong firms is to open up development shops in China. One studio has done this successfully, marrying their core competencies in programming and design in Hong Kong with a large art and animation group in Shenzhen.

3.2. Assessing capability: Four case studies of games companies

To appreciate the issues faced by different types of firms, we will highlight four case studies of games companies with very different products and problems. This will help us to understand the issues and problems relating to the firms' capability and the business environment (especially the constraints on the industry). G Game and F Game are two of the biggest companies. While there are other large games companies, some of them (e.g., Opus) have turned to publishing or distributing games, often those made by developers outside of the country. Apparently, publishing is more lucrative (and easier) than development, as long as the firm has the capital to fund its acquisitions and licensing activities.

3.2.1. M Game – A smaller firm

M Game is a small firm of less than ten employees located in the InnoCenter. They design various kinds of games software including cross-platform (mobile to PC) games as well as an MMO (massively multiplayer, online) game. They used to make mobile games (producing about 15 such games in total), but there is now too much competition in this sector, even though the number of studios has shrunk for the same reason.

The firm was started in part because one of the founders had little to do during the local SARS epidemic. They applied to the science park incubator for funding in 2006. They were trying to survive as a business, in part by trying to produce a "half-product" for another firm. The interviewee noted that it was very difficult to create a product as the market was quite small. They were also apparently too small to go into regional markets or the Chinese market directly. M Game has the capability to develop the software and artistic assets, but lacks a full-time games designer, and other employees share the design tasks. There is a shortage of experienced design talent, but the unwillingness or inability of smaller firms to take on specialized personnel such as designers appears to be an issue. In contrast, mature US firms have specialized lead designers and design teams, using them not only to define the overall game, but to integrate the tasks of the project for other employees. Similarly, as the production values in their products increase, many US firms will employ specialized producers.

M Game has tried to imbue some Hong Kong-specific content and unique ideas into an online game (as part of the action part of the game), which helps their product to stand out. Even the initial stage of production of this online game has been somewhat questionable, as development of these games usually requires a large number of personnel; assigning only half a dozen people to this project will likely involve "cutting corners" (e.g., having less detailed or less complex environments). Therefore, they require more funding in order to produce the game more quickly and to achieve better quality. They cooperate with Japanese clients as well as Microsoft's Xbox platform.

3.2.2. F Game – A niche online computer game company (Japanese-influenced content)

The firm F Game has a a niche market as it was started by afficionados of the Japanese manga comic tradition who were part of a local community who created their own manga. They have just under 20 staff. F Game moved into games from this broader perspective of mixed media. The firm has made one MMO (an action-based MMO like M Game's product), and for this type of game, style (artistic/design) is critical. They have a full-time designer, reflecting a more mature state than smaller companies. Having been in business for several years, they are not eligible for any of the funding available to start-ups. Despite this and the fact that they are located in a warehouse-type building, they have been able to turn out a series of fairly unique products – Japanese anime-inspired games with a slight "non-Japanese" (according to the Japanese) feel to them. They have their hearts in the work, and clearly stay in the business "for the love of the craft." In the past, they dealt with Japanese publishers, and tended to have limited visibility; also they had not made much in the way of revenues, both locally and in Japan. In light of this, at one point they were looking at cross-selling across different media, for example selling the same characters and stories that appear in their games to other media. Recently, they also managed to get funding to produce an online game (a direction that many developers in Hong Kong are moving toward). The budget is significant (about US$1 million) and they needed all of 20 developers to staff the project. F Game recognizes the need to grow bigger, as the company nearly collapsed due to the failure of a game a couple of years ago. They recognize that if they were bigger, they could have multiple projects going at the same time and therefore be able to insulate themselves from a single failure.

3.2.3. G Game – A larger, growing online games company

G Game is an important case which shows how a local firm is able to enter, if not compete, in the Chinese market. They have done this by forming relationships with a leading online games distributor/operator in China, and also by opening a local development arm in China to supplement their resources in Hong Kong. They are developing a massively multiplayer online (MMO) game that appeals to Chinese tastes, partly by relying on Chinese staff to work on the artistic details. Their team size is in-between the larger Chinese studios and the smaller-sized Hong Kong studios. G Game's size appears to be the minimum necessary to break into the more commercially viable segments of the market. There is a regulation that restricts foreign games from entering China, but their current game has been approved by the Chinese government for distribution in China. However, the interviewee (the company's head) did not think it was difficult to get government approval.

3.2.4. D Game Assets – An outsourcing provider for animation in games

D Game Assets, located in Cyberport, is a subsidiary of an Internet company started by a private individual, and of which PCCW is the biggest shareholder. Another sister company, T Games, which does outsourcing work for games, is located on the next floor. D Game Assets started as a web design company, and has since moved to Flash animation, 3-D animation, and game development (mostly the artwork). They often work with their sister company O Technology, which provides the technology. D Game Assets employs 30 people, and with T Games and a Japanese toy company's subsidiary, the total number of employees is 120. O Technology has another 20 employees globally. Like other successful firms, D Game Assets has focused somewhat on IP – in this case, by forming a joint venture with a well-known Japanese toy company, Sanrio, to develop all the digital projects for the *Hello Kitty* franchise. The business model at D Game Assets is to work with "*Hello Kitty*." They mix different Sanrio characters into their digital products. For instance, they are creating the art assets for a *Hello Kitty* massively multiplayer game (Sanrio Digital produces the design, and the programming is outsourced to Taiwan). They also did the design for new *Hello Kitty* services, but these are not games. The company (with Sanrio) targets Asia and the US, and has offices in Korea, Singapore, Japan, and Europe. Surprisingly, 70 percent of users come from the US, 30 percent from SE Asia, and hardly any from Japan (though it was noted in the interview that this could be due to insufficient promotion in the Japanese market by Sanrio). The strategy of D Game Assets is to try to link up with IP owners. At the same time, they provide services to survive, and obtain services from other sister companies as well. Another example of a "safe IP model" is the CGI firm Imagi which acquired or licensed the rights to *Teenage Mutant Ninja Turtles*, *Astroboy*, and *Gatchaman*.

3.3. Analysis: A chasm to cross

One of the problems identified through the case studies is that the smaller firms tend to lack design capability – at the level of the lead designer or design teams in general – as well as the ability to produce highly complex products. This was apparent in the "time sharing" of programmer and artist employees with the design role, and the need or willingness to cut corners. In fact, hardly any of the firms that we interviewed have the resources to compete fully at the highest end of the market. One interviewee (M) noted: "The HK game industry applies a cookie cutter approach by trying to generate similar games with the same engine but different templates." This appears to be an attempt to deal with the limited human and financial resources. To deal with the risks of product "failure," as well as the ebb and flow of human resources over the project's life cycle, firms must be able to have at least two, if not multiple, products in development simultaneously.

A number of other issues were raised during our interviews, including the challenges that Hong Kong firms face. This is evident from Table 11.3, which shows that more mature firms face problems in scaling up financially and resource-wise to tackle the larger projects.

The development of intellectual property, and the creativity needed for conceptualizing that IP or innovative products, does not appear

Table 11.3 Competitive advantages/disadvantages of various firms in the industry/sample

Dimension	Larger, mature companies		Smaller companies	
	F Game	G Game	M Game	Incubatees
Innovative IP	Niche	Incremental	Incremental	Some feasible ideas but little implementation ability
Production expertise	Substantial, higher quality artwork	Substantial. Much of artwork now done in China	Moderate (taking shortcuts to produce games)	Insufficient resources to implement or even prototype ideas
Production value of product	Adequate/good	Adequate/Good	Low	(Production outsourced to other countries)
Other	Financial resources barely sufficient	Sufficient quality to enter Chinese market	Too small to make a major game	Founders from traditional media industries – too inexperienced to make games; insufficient resources

Note: The problems facing firms of differing maturities will vary.

to be a problem for mature firms, as enough Hong Kong firms have demonstrated the creative capability to do so (G Game, F Game, and M Game are all examples of this). The issue is the need to couple this with an increased capability, sometimes read as "production values." Table 11.3 illustrates some of the creative-related features of the different comparators.

For larger firms: The two issues pertaining to capability are: the building of an experienced, skilled team, and the associated need to fund them. While interviewees had suggested that there were sufficient skills in Hong Kong, at least in programming, D Game Assets noted that it was "tricky" to find qualified people in Hong Kong. Programming skills and design were limited. Interviewees pointed out that a large-scale game could not be produced in Hong Kong due to limited human resources. On the other hand, while some people in Hong Kong may have the potential, or the raw talent may be available through the educational institutions, many firms are too resource-constrained to accept or train people in specialized positions such as games designers. Production expertise is also critical, but often overlooked as a requirement for competitive advantage.

The need to finance increased or advanced capabilities (e.g., larger team sizes) is one of the stumbling blocks to building better, more sophisticated games companies. There is general agreement that the venture capitalists in Hong Kong are not familiar enough with games or the other creative industries to participate in this funding. This is not surprising as the same is true of the US and other markets, which rely more on publishers to provide the financing. In China, money has made its way into the sector through various channels, including private individuals, "investments" at the city or government level, and even software firms and firms in other sectors. The larger firms in Hong Kong have somewhere between "barely sufficient" and "sufficient" resources, and those at the "barely sufficient" end just survive from product to product. The firms occasionally derive resources from local and foreign publishers, but publishers often only pay for the development costs and a smaller fixed percentage, so the firms will find it difficult to grow unless they have a breakout hit that they managed to negotiate favorable terms on in advance.

Another problem is that, for those firms that have already gained some accomplishments, in this kind of hits-based industry, even occasional success is not a guarantee of survival. Even the "successful" firms are still too small to survive a single crisis (e.g., product failure). As noted earlier, F Game nearly collapsed when one of its games "failed" in the marketplace. Therefore, there may also be a need to support the already mature or accomplished studios. However, the type of support required by firms needs to be customized and rationalized for their particular purposes. The government's earlier guarantees to the banking sector to support the film industry during its crisis period were not met with great enthusiasm by the banks, which saw the film industry as having high risk and uncertain returns.

For smaller firms: The capability problem is even greater for smaller firms. They are generally too small to have the resources to create a full-fledged product, and neither do they have the requisite experience to do it properly. Not only do smaller firms still have to find financial resources, but without a track record it is difficult for them to attract the financial resources to grow. Another general concern was the lack of experience or resources to create a fully functional product in what is essentially a competitive, mature market. The Cyberport incubatees/start-ups have another problem, which is that while they may have the creative ideas, do not have sufficient experience and resources (if at all) to create a fully fleshed-out product (see Section 4). Many of the incubatees were from the conventional film and CGI industry and lacked experience in the full design and implementation process. The average company founder's profile is said to be that of a more mature industry participant who wanted to start a company. Most are "very creative," and are from traditional media. Most try to write their own (game and other) concepts, and go through many presentation cycles and rounds of publicity, including with foreign publishers (facilitated by the incubator). While some of the ideas may be feasible (according to one interviewee), the harder problem may be how to fully implement the idea in the resource-constrained Hong Kong environment. Most interviewees from outside of the Cyberport incubator had some level of concern about the ability of the incubatees to take their ideas forward. One interviewee observed that one of the incubatees was creating an animated feature "with no hope." It may be that the incubatees could succeed with products that do not have to compete in mature commercial environments with competitors' highly refined products, for example highly niched products in spaces where variety or experimentation is allowed, such as the edutainment sector, or in the case of one former incubatee, one-off projects for public enterprises.

3.4. Other creative industries: Computer-generated imagery (CGI) and 3-D animation

To illustrate the "mixed" nature of Hong Kong's creative industries' capabilities, we will briefly highlight the CGI sector. According to an interviewee, there are four main companies in computer-generated imagery (CGI) or 3-D animation. The four are Imagi (which bought the IP for *Astroboy* and *Teenage Mutant Ninja Turtles* and was already listed on the market), Centro (provider of special effects for *Kill Bill*), Digital Magic (which recently made the animated feature *Yue Fei*), and Menfond (which made the movie *CG 7*), but except for Imagi, they do not own their own intellectual property (IP). These firms are generally large and established, with Centro employing 130 people and Imagi 530 worldwide. The CGI sector is very different from the games sector and has a strong link to the film industry. One interviewee pointed out that CGI firms like Centro, Menfond, and Imagi are all assisting the local film industry. There is less of a linkage between film and games at the moment, although US videogame publishers like Electronic Arts have taken an active interest in using Hollywood intellectual property for content in games. According to the interviewee, there are a lot of incentives that the government could do to bridge gaps. He has visited other (including Western) countries, and found that a lot of companies took advantage of government programs. This view was generally shared by developer and non-developer institutions alike.

4. Is innovating the way to success? An examination of innovation and the "innovation system"

4.1. Small-scale innovation

There are two possible alternatives that firms can play a fundamental part in. The first is to identify low-cost opportunities for innovation. There is still some scope for innovative but smaller-scale products (e.g., casual games) in the area of gameplay, and even for products that might not need substantial resources. However, policies have to recognize these different types of games. No mention of the lower cost/complex casual games "genre" was made in most of our interviews, even though this is a worldwide trend at the moment. For example, one university student start-up in Singapore, Tyler Projects, has monthly revenues in the order of Singapore $50,000 to Singapore $100,000 for their online game – a small-scale Facebook application.

4.2. Own intellectual property creation

A second possibility is that firms could be creative enough to generate their own intellectual property. A large number of firms that were interviewed were developing their own IP. Game developers generally agreed with the importance of IP (or said it was at least implicit in their actions). IP serves a critical role by helping to attract investment, since IP provides firms with better negotiating terms with publishers or other financiers. Having said this, mature firms that were interviewed suggested that the "window" for funding new concepts was past, and that investors generally wanted to see a half-finished or at least a prototyped product. A case in point is the Cyberport's incubatees, many of which offer creative product ideas and are developing original IP. While the incubator seems to have fostered a good focus on IP creation, one non-incubatee interviewee also observed that it was too early for Hong Kong to move into its own IP creation as the market "is young." One company called DCDC had made a CG movie called *Dragonblade* in 2005 which one interviewee said had had a "bad result," and that this "scared off" other firms from trying to develop their own intellectual property.

4.3. Does incubation lead to innovation? Experiences of the Cyberport incubator IncuTrain

It can be argued that the gaming and animation industries in other countries like the US and China are quite independent of institutions such as research labs or universities. Institutions appear to fulfill more basic roles such as the feeding of talent into the industry and the "provision" of founders (i.e., university graduates). We will analyze two advanced initiatives that the Hong Kong government has set up to support or grow the media industry – the incubator at Cyberport and the Hong Kong Polytechnic initiative MERECL (now known as PlayLab).

While the Cyberport itself has been the subject of criticism from various sectors of society, it can be argued that it was run as well as could be expected. It is essential to learn what can be improved in such programs. While there may be a possible advantage to locating all start-ups in a separate "media" hub, or at least under a separate umbrella from science and technology, generally speaking there appears to be less interaction among start-ups in many clusters, for example clusters in the Shijingshan district of Beijing or the IT spin-offs from Singapore. Incubatees have to seek government co-funding or private funding for late stages, but it is also dangerous to assume that the private sector can support start-ups emerging with products of the creative industries that the private sector is unfamiliar with.

Another issue is that, judging from the general experiences of the videogames industry in the US and elsewhere, it would appear that on-the-job experience and numerous "fail cycles" are needed to create the learning necessary for successful companies or their founders. One interviewee observed that the incubator was putting the companies through these learning cycles by way of "presentations." However, a more crucial need is to be able to prototype intensely in order for the product to advance in sophistication over time. For instance, of the three major existing Hong Kong gaming companies (outside of the incubator) that have succeeded (two of which we interviewed), two have previously failed at games production, and only through experience have developed appropriate strategies and product expertise. In media like film, the understanding is that "no one knows" what comprises a hit (Caves, 2000). With videogames, the challenges are compounded by the interactivity of user with the product.

To succeed in this environment, in general, start-ups require access to not just technical knowledge but also production experience, as well as resources, in order to implement their ideas and concepts. My interviews with start-ups in Beijing's Shijingshan district's creative industries cluster suggest that the gaming and animation company founders had this technical and production expertise, along with experience and business acumen or connections. The Cyberport has been trying its best to provide these skills, for example staying at the technological leading edge in console games development by attracting first Microsoft, then Sony, to provide development kits for the incubatees, as well as training and business connections. However, some interviewees noted that the initial degree of rawness of the start-ups was relatively high, and that the management expertise imparted to the Cyberport's incubatees was general and basic in nature. From the brief set of interviews that we conducted, and from the general lack of access to incubatees, it was hard to discern where the experience and expertise required to create "production values" would come from.

This lack of experience may also bedevil another issue, which is the ability to control an outsourcing process. The incubatees are very small in size on average, but a number of them apparently retain the story and design functions, and rely on outsourcing for a lot of their components – for example art assets and programming are done in Eastern Europe and elsewhere. (Console games development cannot be outsourced to China because the country does not have a console games market.)

Finally, the incubator may not be recognizing the reality of commercialization, which is that the typical console project needs considerable investment and experienced personnel. In the order of US$1 million, at a minimum, is needed to begin the development of a concept seriously enough in order to obtain further investments. Several more million dollars and upwards of 30 developers or more are needed to complete a decent console game. On top of this, most incubatees may not have sufficient experience to be able to develop such complex products.

4.4. MERECL's industry-relevant research: Technology push or technology linkages?

One important question to be considered is that of how applied research institutes such as MERECL can usefully contribute to a local creative industry, either in terms of talent, spin-offs, or research. The first potential contribution – that of research – is perhaps the least clear. MERECL's own research projects are fairly innovative, but in the US, academic institutions rarely generate research that immediately becomes part of a development studio's product offerings. If research can contribute anything, it could be in the form of tools or research programs that well-established publishers' internal development studios (which tend to be productivity-oriented) may benefit from. Electronic Arts' support of Carnegie Mellon's Entertainment Technology Center (ETC) is an example of this. One possible reason for this is that most independent development studios do their own core engine (code) development, this being part of their IP as well as a way of creating competitive advantages by linking product features to the code. The second possible benefit is the provision of talent to the industry. However, many researchers who have been trained have not found work locally. Some have moved to other countries, including China, to undertake work, thus defeating one of the stated aims – to train personnel for the local industry. This may have been because the local industry was not large enough or capable enough to use their talents. The third possible benefit is that of spin-offs. This is entirely possible, but again, even in the US, most commercial games are not derived from academic research or research-trained individuals. The sort of enterprise that could be provided would likely be one that has to be tied to a "high-end" user, for example a special effects lab. This very rough analysis suggests that if this sort of research-based gaming technology lab is to benefit local industry more closely, the model and linkages should be rethought some more.

5. Conclusions

The general feeling in Hong Kong is that real estate has been a prime target of private investment and, further, that technology firms and industrialization are not considered to be part of a substantive policy strategy for economic growth. With the prevailing conservative attitudes, it is no wonder that the business environment thus far has been perceived to be stacked against local firms. It has also been observed that the current slate of policies, while somewhat useful for start-ups, are neither sufficient enough to help the start-ups, or geared to help already existing companies that have already "survived" and that are in some cases starting to succeed. Nevertheless, the government and educational institutions have made credible investments in supporting institutions and educational programs. Much learning needs to be done on how these components of the "creative industry system" could better interact with one another. Many interviewees cited the resource limitations facing firms in the games industry, and held up examples of stronger support exhibited by countries such as South Korea and Singapore. The general feeling is that more resources could definitely help the industry – both large and small firms.

One frustration is that current policies such as those operating through the Cyberport are geared to helping weaker, inexperienced entrepreneurs (especially in relation to their "targeted" industries) to start up. Having said that, there may always be a need for creative entrepreneurs, especially if there is a desire for innovative products. For these people, there may be a need to provide opportunities for them to take their creativity through smaller-scale projects, that is, through multiple quick "learning cycles," and not so much through "big" projects such as console games. The casual games business offers opportunities to build smaller versions of what could eventually become more complex products. Programs designed to build up talent in such areas do exist in Singapore as well. There is no equivalent of a fully-fledged media incubator like the Cyberport in Singapore, but many of the pieces do exist to support firms. For instance, there is co-funding for development at early stages, and there are "development" aids such as funding to help firms to license games engines. Some of the lessons from these other programs can be learnt.

In conclusion, it should be noted that as policies toward the new creative industries get formed or reformed, the idiosyncrasies in their products, production processes, and markets must be taken into account. Animation products are made very differently from video games, even though they share one critical input – artists and animators. The way in which users or audiences use or appreciate the products also varies tremendously.

Appendix: List of firms and institutes interviewed (Interviewees from non-game or related sectors not represented)

Table A.1 The study's interview sample and their characteristics

	Subsector	Characteristics	Products	Market
Game studios				
M Game	Game	Small (less than 10)	Online games	HK
F Game	Game	Medium (about 20)	PC games	Japan, Taiwan (limited), HK
G Game	Game	Medium-large (about 10 developers in HK, plus sales; entire art team in China (est. at 15–20))	Online games	HK, China

Table A.1 (Continued)

	Subsector	Characteristics	Products	Market
Other				
D Game Assets (animation assets)	Game	Medium (about 30 locally)	Content for online games etc.	Outsourcing provider (content to (Japanese, etc.)
Z Site (Social Networking Company)	Internet (Social networking)	Small (3–4)	Social networking site	Service provider (HK, Asia)
MERECL Institute	Gaming technology	Medium (about 80–95)		Publicly funded R&D services provider (US)
IncuTrain (Cyberport Incubator)	Support institution for games, animation	47 "graduated" incubatees		Local startups
P Firm Cyberport Incubatee	Social media	Medium	Demonstrate social enterprise applications	HK
Other firms (various)	Design Center, Traditional magazine publisher, IT systems integrator, geographic information software firm			

Notes

1. For the sake of convenience, we refer to the video game, animation, design, computer-generated imagery, and social media sectors as new creative industries (even though some of these have been around for some time).
2. Sources: www.nasscom.com; http:// www.gamasutra.com / php-bin / news_index.php?story= 15529; http://www.gamasutra.com/php-bin/news_index.php?story= 19225.
3. www.virtualworldsnews.com
4. Other non-standard terms are in use. The term *interactive and digital media* (IDM) was coined by the government of Singapore to refer primarily to games, and partly also to animation, but this term is not in wide usage. Another term – new media – tends to refer to the dot-com-influenced web media, so we will continue to refer to the broader set of creative industries, and the entertainment portions of it.

5. http://www.legco.gov.hk/yr08-09/english/panels/itb/papers/itb0209cb1-715-3-e.pdf. Accessed 19 February 2008.
6. http://www.singaporemediafusion.com/news_releases/article043.html. Accessed 14 February 2009.
7. Based on a working paper by F. T. Tschang and J. Comas (available at http://papers.ssrn.com/sol3/papers.cfm?abstract_id=1337188). Accessed 16 February 2009.
8. http://www.salon.com/tech/giga_om/tech_insider/2009/02/15/stop_looking_for_a_wow_killer/Salon.com. Accessed 16 February 2009.
9. http://www.parksassociates.com/research/reports/tocs/2008/gaming2.htm. Salon.com. Accessed 16 February 2009.
10. "Gartner Says 80 Percent of Active Internet Users Will Have a 'Second Life' in the Virtual World by the End of 2011," company press release, 24 April 2007. http://www.gartner.com/it/page.jsp?id= 503861. Accessed 15 February 2009.
11. "Gartner Says 90 Percent of Corporate Virtual World Projects Fail Within 18 Months," Gartner Research, Press Release, 15 May 2008, http://www.gartner.com/it/page.jsp?id=670507. Accessed 11 June 2010.
12. According to a LegCo report: http://www.legco.gov.hk/yr08-09/english/panels/ci/papers/ci1118cb1-201-6-e.pdf. Accessed 19 February 2009. Another source puts the number of incubatees at 47: LC Paper No. CB(1)201/08-09(06), 18 November 2008, Legislative Council, Hong Kong Special Administrative Region of the People's Republic of China, Legislative Council Panel on Commerce and Industry, Update on Cyberport Digital Entertainment Incubation-cum-Training Programme.
13. Partly based on IDC (2007).
14. http://www.legco.gov.hk/yr08-09/english/panels/ci/papers/ci1118cb1-201-6-e.pdf. Accessed 19 February 2009.

References

Aoyama, Y. and H. Izushi (2003) "Hardware gimmick or cultural innovation? Technological, cultural, and social foundations of the Japanese video game industry," *Research Policy*, 32(3), 423–444, March.

Arora, A. and A. Gambardella (2005) *From Underdogs to Tigers: The Rise and Growth of the Software Industry in Brazil, China, India, Ireland and Israel*, A. Arora and A. Gambardella (eds), New York, Oxford University Press.

Caves, R. (2000) *Creative Industries: Contracts between Art and Commerce*. Cambridge, Mass.: Harvard University Press.

DCMS (2001) *Creative Industries Mapping Document*. Department of Culture, Media and Sport, United Kingdom.

Entertainment Software Association (2008) *2008 Essential Facts About the Computer and Video Game Industry*. Washington, D.C.: Entertainment Software Association.

IDC (2007) IDC Hong Kong Digital Entertainment Study (2006–2007) Executive Summary, IDC Corporation.

Teo, Stephen (1997) *Hong Kong Cinema: The Extra Dimensions*. London: British Film Institute.

Toh, M. H., A. Choo, T. Ho, Economics Division, Ministry of Trade and Industry (2003) Economic Contributions of Singapore's Creative Industries, Economic Survey of Singapore, First Quarter. http://app.mica.gov.sg/Data/0/PDF/6_MTI%20Creative%20Industries.pdf. Accessed 17 February 2009.

Tschang, F. T. (2007) "Balancing the tensions between rationalization and creativity in the video games industry," *Organization Science*, 18(6), 989–1005.

Tschang, F. T (2005) "Videogames as interactive experiential products and their manner of development," *International Journal of Innovation Management*, 9(1), 103–131.

Tschang, F. T. and S. Tsang (2008) China's New Media Sectors: Domestic Culture as Competitive Advantage, *Innovation in Greater China*, H. S. Rowen, M. G. Hancock and W. F. Miller, eds, Shorenstein APARC (center), Stanford, CA.: Stanford University.

www.17173.com (2007) China Online Games Investigative Report 2007.

12
Environmental Technology: Hong Kong's Innovation System

Atsushi Sunami
Director, Science and Technology Policy Program
Associate Professor
National Graduate Institute for Policy Studies

1. Introduction

Hong Kong is going through an era of rapid transformation as it tries to remain as one of the leading metropolitan centers of Asia at least for the next several decades. While China's economy is expanding rapidly, Hong Kong has served the role of the gateway for channeling foreign investment to Mainland China. As Hong Kong competes with the major cities in the coastal area of the Mainland, it needs to transform itself from a global financial service center as well as a trade support center for Asia, to an important access point in the global innovation network. In today's global network for knowledge creation, Hong Kong can play an important role in trading and importing technology and jointly developing technology for the local Chinese market. Especially, the collaboration with Guangzhou and Shenzhen will be the essential aspect of Hong Kong's future economic development.

Currently, China is searching for a new strategy of sustainable economic development. The Chinese leadership considers the development of "green technology" as essential for transforming its current mode of economic development into a more energy-efficient and environmentally friendly socio-economic system.[1] In the past decade, many Hong Kong manufacturers have relocated their factories to Guangzhou and the Pearl River Delta (PRD) region. There are over 56,000 Hong Kong factories currently located in that region. Consequently, these factories are seeking to adopt better and more affordable technology for cleaning up their waste water and air in order to meet environmental standards.

In Mainland China, a number of new and also amended environmental laws and regulations were passed in 2007. Among them, the laws targeting energy conservation, water pollution, environmental information disclosure, and the ban on plastic bags are well recognized, even outside China. As China pushes for tougher environmental regulations, the heart of China's manufacturing base in the south creates a potentially huge market for environmental technology. All of the technologies needs to be modified or redeveloped to meet the local need before they are adopted. Moreover, even after it has been successfully adopted, the system needs to be monitored regularly and readjusted to a constantly changing environment. Here, Hong Kong can potentially be a significant player in developing the technology to be adapted more easily to the local factories using more advanced technology in the US, Europe, and Japan. There are two other, but important, reasons why Hong Kong should help China, especially Guangdong, to clean up the environment. One is the deterioration of Hong Kong's air quality, primarily caused by the air pollution from the PRD. The other is the possible rise of the sea level due to global warming. Both would threaten the future welfare of Hong Kong's living conditions.

Environmental technology is a very broad area of technology, ranging from a simple energy-saving device to a complicated system of water purification and recycling. Environmental technology can be conveniently divided into four areas based on their application: (1) global warming, (2) environmental pollution, (3) resource recycling, and (4) ecological management. The main users of the technology are business and public sectors. Therefore, it is fair to say that the environmental technology developed by Hong Kong is mainly used by the local governments and local businesses of Hong Kong and the PRD region. Although there is no definite boundary for the components of environmental technology, it can be said to encompass all technology areas that are used to maintain the sustainability of the global green growth. It means the technology that helps to improve our environment, the technology that measures and evaluates the current environmental conditions and potential risks. It also means the technology that leads to alternative sources of energy, and the technology that resolves or improves an already deteriorated environment. However, this chapter will focus mainly on the technology area relating to environmental pollution. Hong Kong can take this system of technology not only to the local market but also for application in Mainland China.[2]

There are four basic suggestions for policy action one could consider when looking at the future of environmental technology in Hong Kong.

(1) The Hong Kong government's role in creating the demand for environmental technology can have a significant demand-pull effect on the development of the technology. (2) The collaboration with Mainland China, especially Guangdong local government, is a necessary condition for bringing Hong Kong's environmental technology to the PRD region. (3) China and Hong Kong's commitment to solving climate change creates a great opportunity for Hong Kong's environmental technology development. (4) In order to sustain the development of innovative capabilities, Hong Kong should implement the innovation strategy based on creating a system utilizing the capability of universities through designing the channels of technology transfer to the local industries in particular. Hong Kong's innovation capabilities have been well recognized. According to the Global Innovation Index compiled by Confederation of Indian Industry (CII) and INSEAD, Hong Kong ranked 12th in the GII 2008/2009 overall rankings following Singapore, Korea, and Japan among the Asian countries.[3] In addition, Hong Kong universities maintain quite a strong performance in scientific research while continuing to attract good students from the Mainland.

2. Hong Kong's environmental problem

The deterioration of Hong Kong's air quality has become one of the city's main policy concerns in recent years. Although emissions of key air pollutants in Hong Kong have dropped substantially since the 1990s, the visibility has deteriorated significantly, reflecting a worsening of the regional smog originating from the PRD in China (see Figure 12.1).

Two distinct sources of pollution are local pollution sources (see Table 12.1) and regional pollution sources coming from the PRD region. The pie charts in Figure 12.2 clearly show that the main source of pollution comes from the PRD region.

Hong Kong can become one of the important hubs of the global innovation network for the development of environmental technology, transforming from the gateway of China for foreign direct investment (FDI). By examining the current activities relating to environmental technology, especially for wastewater management and air pollution control in Hong Kong and the PRD region, it will show that already there is a basic trend toward Hong Kong becoming an active and important player in bridging the world technology and the local market. Furthermore, it will shed light on the fact that if the public sector and

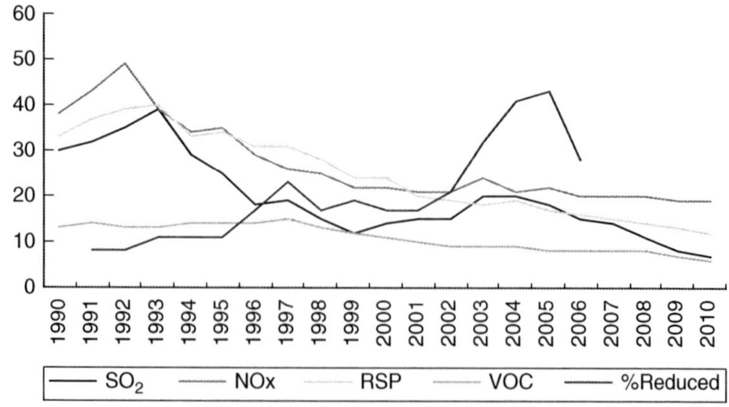

Figure 12.1 Air pollutants emitted and visibility trend from 1990 to 2010
Notes: The rate of reduced visibility refers to the percentage of time in a year which has a visibility of less than 8 km, with relative humidity not exceeding 80 percent. The emission figures after 2006 are projected data.
Source: Hong Kong Department of Environmental Protection.

Table 12.1 Local pollution sources

Electricity generation	**Road transport**
Largest source of SO_2, NOx, and RSP emissions.	Second largest source of NOx, RSP, and VOC emissions.
Navigation	**Civil aviation**
Second largest source of SO_2 emissions.	A relatively minor source of emissions.
Other fuel combustion	**Non-combustion sources**
A relatively large source of RSP emissions (from fuels consumed in industrial, commercial, and residential sectors, off-road transport, construction industry, etc.).	Largest source of VOC emissions (from consumer products, printing, paints, etc.).

Source: Hong Kong Department of Environmental Protection.

a series of adequate public policies create a demand-pull effect for the development of pollution control and alternative energy technology, Hong Kong can make use of its full potential to be once again the gateway and a leading metropolitan center in Asia for the global network of innovation in "green technology."

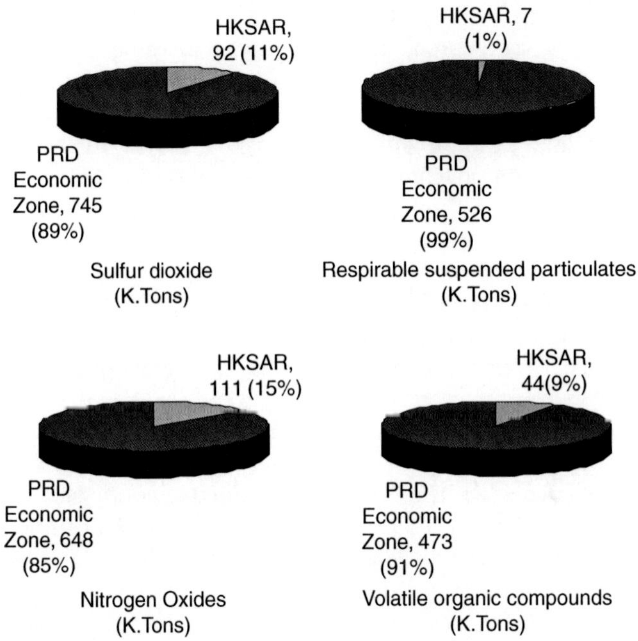

Figure 12.2 Emissions in the PRD region in 2003
Source: Hong Kong Department of Environmental Protection.

Hong Kong already has several institutions actively engaged in the development of environmental technology. On the one hand, most of the fundamental research has been carried out by the universities. On the other hand, the government-affiliated institution, the Hong Kong Productivity Council (the Productivity Council or HKPC), is actively developing the prototype technology and diffusing that technology through local businesses in the region. First, the HKPC's environmental technology program will be discussed, followed by other institutions engaged in the development and the diffusion of "green tech" in the region, namely, Hong Kong R&D Centers, Hong Kong Science Park, and universities.

3. Institutions

3.1. Hong Kong Productivity Council's "green productivity" program

Hong Kong Productivity Council (the Productivity Council or HKPC) was established almost 40 years ago to provide services supporting local business in various aspects, with a focus on manufacturing in Hong Kong. It was a part of the "productivity movement" in the whole Asian region driven by the Japanese business community in their effort to help economic development in Asia through promoting efficiency in manufacturing and improving productivity. Currently, the Asia Productivity Organization, an international organization located in Japan, is continuing this effort through its extensive network of Asian productivity organizations spread across over the whole of Asia. The technology services they provide range from manufacturing technologies and information technologies, to business management and training as well as environmental technologies more recently. The Productivity Council has developed an extensive network among the Hong Kong manufacturing industry over the years. The network is the strength of the Council especially in the diffusion of environmental technology among the PRD factories in particular.

The HKPC launched the first environmental support service for the business community in Hong Kong as early as 1981 by opening its laboratory testing activity. In fact, the "green productivity" concept was presented by the Productivity Council in Hong Kong. Since then, the Productivity Council has pushed the "green productivity" movement for this region as part of the Asia Productivity Organization's region-wide effort to help the development of the environment technology industry. In particular, the need for environmental support services, both in technology development as well as in the application and implementation of the technology, is prominent in the PRD, where many Hong Kong's manufacturing firms are located or have contracts with factories in the region.

Today, the PRD region is known not only as the manufacturing center of South China, but also for the rapid deterioration of its environment. Over the years, serious pollution caused by these factories has deteriorated the quality of air and water. As China pushes hard for the cleaning up of the environment based on the current 11th 5-year Plan, among other central government programs, the PRD's need for effective environmental support services will increase even more significantly in the coming years.[4]

Under the Hong Kong Special Administrative Region (HKSAR) government's "Action Blue Sky Campaign," HKPC's push for "green productivity" is carried out in accordance with the policy of both the Hong Kong and Beijing governments. Furthermore, it tries to take a lead in meeting various international environmental standards including the EU's, some of the toughest regulations, such as WEEE, RoHS, and EuP.[5]

Roadmap of the "Green Productivity" movement:

1981 – Launch of laboratory services
1983 – Launch of services in pollution control for the tanning and leather dressing industry as well as the electroplating industry
1984 – Building of the first flue gas desulfurization system in Hong Kong
1985 – Launch of technical support services in pollution control for SMEs
1986 – Establishment of the Center of Environmental Technology for Industry, design of locally fabricated cost-effective pollution control systems
1988 – Launch of Indoor Air Quality services
1989 – Launch of supporting services for infrastructure development projects
1992 – Launch of environmental compliance supporting services for Hong Kong-linked factories
1993 – Development of cleaner production technology
1994 – Introduction of ISO 14000 series Environmental Management Standard
1997 – Introduction of "Energy Performance Contracting" services
2005 – Launch of Green Manufacturing Supporting Services: Design for Environment, EU RoHS/WEEE compliance, EuP compliance, China RoHS
(From Hong kong Productivity Council, "Green Productivity, 25 years and beyond").

Some examples of the commercialization of environmental technology developed through the support provided by HKPC are as follows: Handy Toilet Waste Disinfection System, Wastewater Treatment System for Construction Sites, Automatic Wheel Tire Washing System, Secondary Treatment System for temporary offices on construction sites.

The top priority for the Productivity Council's green-tech projects is the wastewater treatment in the PRD region. Almost 70–80 percent of the Hong Kong's industry's operations are taking place in the region. Moreover, up to 70,000 Hong Kong-owned enterprises have relocated to the Mainland. Therefore, water-treatment and water-recycling technology is much needed in the region, especially as the Chinese authorities have tightened environmental regulations in the last few years.[6]

The Productivity Council has set up a one-stop service for local factories from R&D, design to the system implementation. Since most of the PRD region's projects are small, and individual factory-based, the time line is usually very short and therefore quick and cost-efficient solutions to the problems are required. Overseas technologies are usually more advanced than the locally available ones.[7] However, they tend to be too expensive for the local market. The Productivity Council's strategy is to import the core technology from Germany or Japan and to modify and assemble it using OEMs (original equipment manufacturers) in the region. In short, the role of the Productivity Council is to act as a bridge between the more advanced overseas technologies and the local needs by helping to localize the technology. This is exactly the strategy Hong Kong should adopt for the development of environmental technology in the initial phase. HKPC has successfully led many projects in the region. For air quality improvement, it has developed and installed air filtering and control systems for the metal product industry, the food industry, and for automobile repair workshops. In the area of wastewater treatment systems, HKPC has helped companies in the region's electronics industry, toy manufacturing, chemical industry, and food industry, including multinational corporations (MNCs). Recently, a Japanese company in the region used HKPC's wastewater treatment system to meet the regulatory requirements instead of using more advanced Japanese technology, because of the high cost of the latter and the need for further modifications of that technology to fit the local environment. The technology that HKPC has introduced for wastewater treatment combines both a biological submerged aerated filter (SAF) and a rotating biological contactor (RBC) for the food industry at a reasonable cost.

However, the challenge for the operations in the PRD region is that the environmental regulations are often not so clear when they are implemented and changed very often. These include various standards or quotas for different organic pollutants. Therefore, a constant collaboration between the central and regional governments is crucial. This is also an advantage for Hong Kong as it has a good working relationship with the local governments in the region compared with MNCs operating in China. In fact, there are several MNCs who have partnered with the Productivity Council to bring the technology into the region.

3.2. Hong Kong Research and Development Centers

3.2.1. NAMI at HKUST

The R&D Center program of Hong Kong has been set up by the HKSAR government to facilitate the collaborative R&D between the research organizations and industry. Under this program, six R&D Centers have been established so far. However, several of these centers suffer from inadequate financial support from the government, while operating under the constant demand for short-term results for evaluation. Therefore, the scale of the activities of these R&D Centers is comparatively small for competing at world level or even when compared with most of the R&D Centers in Mainland China.[8]

The Nanotechnology and Advanced Materials Institute (NAMI) R&D Centre is located in the Hong Kong University of Science and Technology (HKUST). The center's research focus is in the following five key areas: nanomaterials, nano-opto-electronics, nano-structured/textured materials applications, advanced materials for interconnection and packaging, and development of advanced materials.[9] However, because of the lack of adequate funding and support from the government as well as from the university, the center operates rather randomly in these areas. The limited resources have forced the center to abandon some research areas and focus on more near-sighted research topics with clear industrial support. Here, the center is working on the technology applicable to the needs of the PRD region's manufacturers in meeting environmental regulations, rather than focusing on more fundamental research. In particular, the center is pushing hard for the project on wastewater management with photocatalystic and new materials technology. The director, who has many years' experience of working with the top chemical companies, regularly visits the factories in the PRD region to oversee the project. Since these projects are still in the initial stage, there are no clear results to be seen at present. Furthermore, the center's network with the PRD manufacturers is still at a personal level and not fully institutionalized. Moreover, the director recognizes the presence of several MNCs in PRD already, which will inevitably become tough competitors for the center in providing environmental technology.

At any case, it is essential for NAMI to establish a role in developing the technology for the PRD quickly, as many MNCs are seeking an opportunity to scale up their activities in the region. Furthermore, in order to compensate for its limited resources (less than ten principal investigator (PI) level R&D staff), NAMI must find a way to expand its collaborative R&D network with the advanced R&D Centers outside Hong Kong. Consequently, the center has recently begun to look for partners in the US, EU, and Japan.

3.3. Hong Kong Science and Technology Park: Incu-Tech

The Hong Kong Science and Technology Park (HKSTP) has been operating an incubation program called Incu-Tech. The incubation program provides the services necessary for start-ups, including consultancy services, entrepreneurship training, and networking with local universities and the other research centers, both in Hong Kong and elsewhere. All tenants are eligible for rent-free space in the first year, followed by a 2-year term with reduced rent. The main areas of technology are IT, electronics, biotech, and precision engineering. Although, it has not made a significant contribution, there are two or three start-ups based on environmental technology that have already graduated from the program. The HKSTP is not currently interested in pushing for the development of environmental technology as they will be focusing on IT and precision technology for at least next few years. One of the companies in the HKSTP is AKOS (Advanced Technology Ltd). Their main product is an air purifier using basic technology from Germany. AKOS is trying to market their products mainly to meet local Hong Kong household needs. Other companies that have developed as spin-offs from local universities or by using technology developed by the university include Environmental Care (CU) and Altenano (HKUST). All of these companies are manufacturing the prototypes by contracting out to the PRD factories. Both Environmental Care and Altenano are aiming at the local market and the PRD region as well as the global market.

3.4. Universities and U-I linkages

Hong Kong's strength in developing environmental technology comes from basic research undertaken in the local university labs. In particular, chemistry is a comparatively strong area of fundamental research in the Hong Kong academies. Based on the data provided by Nature China, Hong Kong is clearly one of the top research centers in China (see Table 12.2). In particular, the University of Hong Kong and HKUST are both ranked in the top ten institutions producing high-quality research papers accepted by Nature China (see Table 12.3). Also, China's relative strength lies in the field of chemistry, as indicated by the number of high citation papers (see Figure 12.3).

A good example to illustrate how Hong Kong has a great potential to excel in several areas of science by utilizing the right resources available is Professor Jimmy Yu at the Chinese University of Hong Kong. His research in Advanced Photocatalyst and Nano Coating led to the award of Hong Kong's Innovation Technology Funding in 2001, in collaboration with local company, Environmental Care Ltd.[10] As in

Table 12.2 Top 10 Chinese Institutes Producing the Articles Featured on *Nature China* 2007–2008

City	Number of Papers	% of Total
Beijing	153	30%
Shanghai	99	19%
Hong Kong	60	12%
Nanjing	26	5%
Hefei	25	5%
Guangzhou	18	4%
Wuhan	13	3%
Hangzhou	11	2%
Lanzhou	11	2%
Xiamen	11	2%

Source: Data provided by Nature China.

Table 12.3 Top 10 Chinese Institutes Producing the Articles Featured on *Nature China* 2007–2008

Lead Institute	# of Papers	% of Total
Chinese Academy of Sciences, Beijing, China	60	13%
Chinese Academy of Sciences, Shanghai, China	50	10%
University of Hong Kong, Hong Kong, China	36	8%
Peking University, Beijing, China	29	6%
Tsinghua University, Beijing, China	21	4%
Fudan University, Shanghai, China	16	3%
Hong Kong University of Science and Technology, Hong Kong, China	16	3%
University of Science and Technology of China, Hefei, China	16	3%
Shanghai Jiao Tong University, Shanghai, China	11	2%
Zhejiang University, Hangzhou, China	10	2%

Source: Data provided by Nature China.

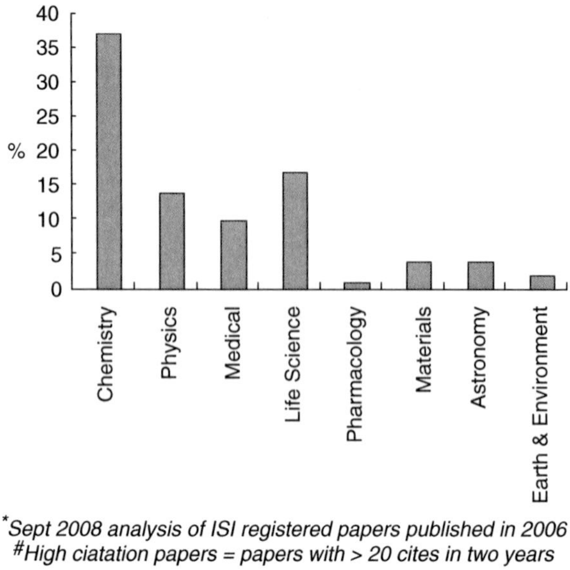

Figure 12.3 % of High Citation Papers by Field
Source: Data Presented by Felix Cheung, Nature China at JST-CRC Symposium, 2008.

other notable cases of university–industry collaboration, Professor Yu met representatives from Environmental Care at an academic conference almost by accident. He did have some interest in bringing his research to the market by partnering with industry to develop certain products using his technology. However, it was the people from Environmental Care who approached him at the conference without knowing his interest.

Environmental Care used to be a ship building company but then diversified into environmental technology. The company and Professor Yu shared a mutual interest in taking the professor's research into the market. Environmental Care's factories for developing their technology are in Shenzhen. They produce salt water purifiers, currently installed at a famous seafood restaurant in Hong Kong.

Hong Kong's strength in environmental technology lies in (1) information by tapping into its global network, (2) intellectual property, and (3) international collaboration. Furthermore, the availability of human resources in S&T from Mainland China is a new addition to the list. For example, Professor Yu's students are mostly from the Mainland. The strength of China's science presently lies in chemistry, math, and physics. English is the working language in Hong Kong's universities;

however, Chinese can be used outside of campus, which is quite helpful especially for the Mainland students.

On the other side, the problem of research funding is still a significant constraint for Hong Kong's innovation system. To remedy the problem, the Research Grant Council is providing funds of HK$18 million. However, the R&D projects have been constantly under pressure for review and evaluation by the government. This affects research directions and favors more short-term research targets rather than the seamless research and development efforts necessary for technological development. As for Hong Kong as a potential market for green technology, environmentally friendly construction or other forms of energy saving for the consumer sector can be considered in the future. However, the local power company is very slow in adopting renewable energy or new efficient technology. Moreover, the private sector is still only looking at short-term targets.

Technology transfer from university to industry is rather difficult and not working as well in Hong Kong as in other major advanced economies such as United States. Thus, the problem of Hong Kong's environmental technology is how to get the technology out into the market. The process of technology transfer and more applied research by industry are the critical issues facing Hong Kong's innovation system in general.[11] In order to strengthen the role of the industrial sector in promoting environmental technology, the Hong Kong Environmental Industry Association has been formed, led by several leading companies in Hong Kong such as Dunwell Enviro-Tech Ltd. The activities of the association should be recognized as an important step in transforming Hong Kong's industry and, supported by the government, through public private partnership.

4. Evaluating Hong Kong's innovative capabilities in environmental technology

It is quite difficult to say exactly how well Hong Kong's innovative capability competes with the rest of the world, especially the US, Europe, and Japan, in the field of environmental technology.

As it has indicated earlier, environmental technology covers a very broad area of technology. For convenience, environmental technology can be divided into four areas based on their application. These are technologies related to (1) global warming, (2) environmental pollution, (3) resource recycling, and (4) ecological management. In the earlier part of this chapter, technology related to cleaning up environmental pollution, both air and water, has been discussed extensively, as Hong Kong hopes to sell this technology first to Mainland China's factories. Japan Science and Technology Agency's Center for Research and Development Strategy conducted a detailed survey of environmental technology in the US, Europe, China, Korea, and Japan in 2008. This comparative study of five countries used the above four categories to cover the whole field of environmental technology. It then introduced ten subfields of technology within the four categories to break down the subject further. The study looked at these subfields of environmental technology in the US, Europe, Korea, China, and Japan from three perspectives: (1) the level of basic research, (2) the level of technological development, and (3) the level of commercialization. According to the study, in almost all fields of technology, the US, Europe, and Japan excel in the levels of basic research, development, and commercialization of these technologies.[12] In short, it is clear that the advanced industrial economies are still the leaders in environmental technology.

Environmental technology is a product of multidisciplinary research cutting across several areas of traditional research disciplines. Thus, research collaboration among different disciplines and institutions becomes a key to the success of the development of the technology. Several studies on centers of excellence, such as the reports by the Science Policy Research Unit of Sussex University and National Institute of Science and Technology Policy of Japan, show the existence of a global network of research activities among the leading research centers of environmental technology in the world.[13] The above mentioned Japan Science Technology Agency study also shows a dramatic increase in the level of research in the area of biodiversity in China due to the effect of brain circulation. The number of Chinese researchers who have returned to China from overseas are continuing to do research with leading scientists in the US and the Europe. This has helped to bring up the level of research in China in those fields.

However, it is also true that the level of technological advance is not as important as the choice of technology as long as it delivers the required results. In other words, the advancement of environmental technology depends not only on innovative capability but also on how much the technology is in fact used. For example, according to the JST study Japan's recycling technology for building materials is leading the world because of Japan's strict laws and its enforcement of the building recycling code. Therefore, it is as important to see the technology as user-driven as it is to consider its innovation capability when considering the strategy to adopt in developing environmental technology for Hong Kong.

5. Suggestions for policy action

5.1. The Hong Kong government's policy for creating the demand for environmental technology: A demand-pull strategy

It is now becoming clear that the development of Hong Kong's environmental technology industry is closely related to its presence in the PRD region. Thus, it is very important to investigate the effectiveness of Hong Kong's role as both mediator and modification development center for the introduction of environmental technology in the PRD. In fact, there is always the possibility that the PRD can develop its own environmental technology industry without Hong Kong by working directly with Beijing and MNCs.

The other possibility is the environmental technology applied in the local Hong Kong market, one example being the construction industry. However, many of Hong Kong's real estate developers are known for being very cost conscious and interested more in short-term projects. Tougher building regulations and a change in consumer taste might create an opportunity for the development of environmental technology in areas such as energy efficiency and building materials recycling. It might be useful to discuss this possibility with those working in the construction sector and real estate development as well as with the government in charge of environmental regulation in this area.

Hong Kong has little local environmental technology available. The main reason is the lack of demand in the market. The Environment Bureau was set up in July 2007 as part of the reorganization of the Government Secretariat with the commencement of the Third Team of the Hong Kong SAR government. The policy objectives of the Environmental Bureau are environmental protection, green energy, and sustainable development.

In 2007, Donald Tsang, in his "New Direction for Hong Kong" address, announced his commitment to improving the environment. This shows potentially that the market for green technology will expand in Hong Kong in the next few years. The current measures for controlling air pollution are (1) new coal-fired power plants banned since 1997 and (2) tightening of emission caps. New measures introduced recently are (1) legislative amendments to stipulate the emission caps for 2010 and beyond, and to enable power plants to conduct emissions trading, (2) link power companies' rate of return to achievement of the emission caps, and (3) provide higher return to renewable energy facilities. Regarding electricity generation and demand-side management, existing measures include (1) promotion of energy efficiency and conservation and (2) enhancement of building efficiency. New measures under public consultation include (1) mandatory compliance with the Building Energy Codes and (2) mandatory Energy Efficiency Labeling Scheme.

The existing measures for environmental regulations regarding transportation are (1) a world-class public transportation system consisting of LPG taxis and Light buses, (2) tightening of emission standard to Euro IV, (3) requirement for pre-Euro diesel vehicles to install emission reduction devices, (4) provision of HK$3.2 billion to expedite replacement of pre-Euro and Euro I commercial diesel vehicles with Euro IV models, and (5) provision of incentives through tax breaks for vehicles with low emissions and high fuel efficiency. The new measures are (1) to ask the public to enact a ban on idling cars with running engines, (2) to provide tax incentives to encourage the use of the Euro V diesel, and (3) to strengthen the control of emissions from petrol and LPG vehicles by the use of remote sensing equipment and advanced emissions testing.

With regard to the policy concerning the area of navigation, the existing measure has required the usage of ultra-low-sulfur diesel in government vessels since 2001. The new measures are (1) to study the feasibility of requiring all vessels plying the harbor to use high-quality fuel, (2) to explore with Guangdong on more stringent measures to control emissions from vessels in PRD for inclusion in the Regional Air Quality Management Plan, and (3) to participate in the International Maritime Organization and support the development of emission control measures for ports and ocean liners.

Other important existing measures are (1) to require a vapor recovery system for vehicle refueling to be installed in all petrol stations and (2) to impose mandatory emissions limits for volatile organic compounds in architectural paints, printing ink, and selected consumer products. The new measures are (1) to mandate the use of ultra-low-sulfur diesel in all industrial and commercial processes, (2) to enhance public awareness and education, for example encourage the business sector to internalize core environmental values in investment decisions/policies,

(3) to require appropriate environmental performance in procurement and merchandizing activities, (4) to promote clean production, and (5) to launch a HK$93 million project to provide support for Hong Kong-owned operations in the PRD for cleaner technologies and practices. The government must provide a firm political leadership and commitment in implementing these already existing measures as well as following them up with various new initiatives to improve the environment of air and water as well as recycling/reusing resources for Hong Kong.

5.2. Collaborating with Guangdong government

The governments of Hong Kong and Guangdong agreed to reduce the emissions from the following sources of air pollution by 2010 (compared with 1997) in April 2002: (1) sulfur dioxide (SO_2) by 40 percent, (2) nitrogen oxides (NOx) by 20 percent, (3) respirable suspended particulates (RSP) by 55 percent, and (4) volatile organic compounds (VOC) by 55 percent. The two parties are developing the Joint Air Quality Monitoring Network to oversee the regulations.

The environmental regulations and measures currently existing in the PRD are (1) to reduce emissions from power generation, (2) to continue to reduce energy intensity (i.e., energy consumed for each CNY 10,000 GDP): Guangdong Province: −16 percent by 2010 and PRD

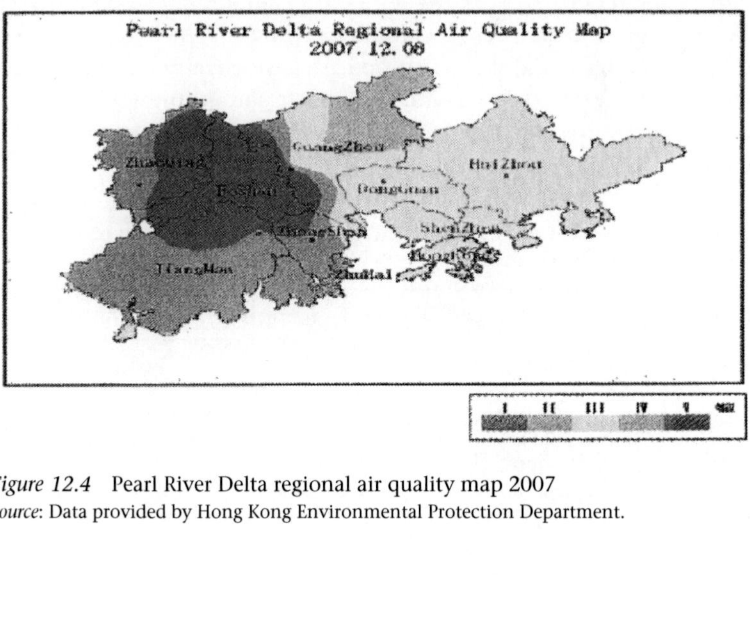

Figure 12.4 Pearl River Delta regional air quality map 2007
Source: Data provided by Hong Kong Environmental Protection Department.

Economic Zone: −18 percent by 2010, (3) to install gas de-sulfurization facilities in coal-fired power plants, (4) to phase out highly polluting small thermal power units, small-scale cement factories, and iron and steel mills, and (5) to develop new large-scale gas-fired power plants. Moreover, major regulations for transportation include (1) to reduce vehicle emissions, (2) to adopt the National III (on a par with Euro III) motor vehicle emission standards and motor fuels, (3) to reduce emissions from the industrial sector, (4) to phase out highly polluting industrial processes and bring in compulsory adoption of cleaner production practices for some industries, and (5) to control emissions of VOC from oil depots, tanker trucks, and petrol filling stations.

Furthermore, Guangdong has introduced measures to meet the 2010 emissions reduction target (see Table 12.4). These are: (1) new coal-fired power plants to install de-NOx equipment, (2) more stringent emission standards for boilers, (3) tighter emission control on local vessels, (4) cleaner production requirements for paint production, (5) green industrial standards for the printing industry, (6) labeling schemes for VOC-containing products, and (7) a public awareness program to promote the use of products with a low VOC content. The challenge effectively is to implement all of these measures and monitor the progress correctly. This requires collaboration between the two governments to build up an effective governing mechanism for environmental regulation in the region, such as the Joint Air Quality Monitoring Network. The two governments must continue to work together to address the common problem and should introduce a joint government program to build the PRD region as a regional platform for an innovation ecosystem targeting the development of environmental technology for the region.

Table 12.4 Projected emissions in 2010 after the implementation of additional control measures by Guangdong

Pollutant	**2010 emission reduction (using 1997 as base year)**		
	HKSAR	**PRD Economic Zone**	**Reduction target**
SO_2	−54%	−41%	−40%
NOx	−25%	−20%	−20%
PM_{10}	−58%	−60%	−55%
VOC	−55%	−56%	−55%

Source: Hong Kong Environmental Protection Department.

5.3. Climate change

The situation of Hong Kong with respect to the problem of climate change is as follows. The local greenhouse gas emissions are relatively low at some 45,000 Kt. CO_2 equivalent per annum. The per capita emission is 6.5 tons CO_2 equivalent, and the carbon intensity is 27.6 kg CO_2 equivalent per HK$1000 GDP (2005). The main emission sources in Hong Kong for greenhouse gas are the energy sector (63 percent), transport sector (16 percent), and waste/landfill (12 percent), according to the Environmental Protection Development, Hong Kong. The greenhouse gas reduction measures mainly adopted are (1) electricity generation, (2) renewable energy, (3) efficient public transport, (4) landfill gas utilization, (5) energy efficiency and conservation, and (6) publicity and education. Furthermore, other additional measures being considered include: to consult the public on the mandatory implementation of Building Energy Codes, to launch the mandatory Energy Efficiency Labeling Scheme, to conduct a carbon audit and implement an emissions reduction campaign at the new central government complex at Tamar, to incorporate environmental measures under the New Scheme of Control Agreements, to encourage the use of biofuel in vehicles, and to conduct a climate change consultancy study.

With regard to the efficiency in production of energy and the development of alternative energy, the Energy Division of the Environment Bureau is responsible for formulating the energy policy and overseeing the economic regulation of the energy sector. Safety regulation is overseen by the Electrical and Mechanical Services Department, while environmental regulation is enforced by the Environmental Protection Department.

The objectives of energy policy are to ensure reliable, safe, and efficient energy supplies at reasonable prices, and to minimize the environmental impact caused by the production and use of energy. The main measures are to achieve the energy policy objectives via legislation, "agreements" guidelines, and public education, etc. The monitoring mechanism is as follows: supply is by the private sector, with no subsidy from the government, setting up safety and environmental standards via legislation and guidelines. For town gas, the policy instrument is based on an Information and Consultation Agreement. For the fuel oil market, free market economic principles are used as the main policy tool to encourage competition and to enhance transparency. Consequently, if most of these policy measures are implemented effectively by a strong political leadership, Hong Kong can create sufficient local market for environmental technology to be developed even further as part of its demand-pull strategy.

5.4. Innovation strategy

The innovation policies for environmental technology in most countries share several common features: funding, the institutions supporting networks of technology, IP/standard, public procurement, human resource development, and the global network of talents. Hong Kong does have some advantage in creating the global networks for technologies as well as the human capital to enable it to be a regional platform.

The success of the HKPC's initiative in developing an innovative environmental technology industry in Hong Kong and the PRD lies in several factors. The first of these is coordination with the regulatory agencies responsible for implementing environmental policy in order to guide the direction of technology development as well as identify the adequate levels and forms of green technology for the region. Second, the role of the local government, not only as a regulatory agent but also as an important user of environmental technology, should be recognized. In short, how effective Hong Kong is as a mediator in adopting and implementing environmental technology in the region will be the key for the development of the new industry. Third, given the role of Hong Kong as the gateway for the PRD region, it can collaborate with the advanced economies such as the US, Japan, and the EU in R&D for any environmental technology used in the region. In particular, air pollution control, wastewater management, and energy and material recycling/reuse (waste chemicals) systems are the main areas of environmental technology. The source of the fundamental technology can come from the advanced economies, while Hong Kong can provide the necessary modifications to meet the local needs and be cost-effective.

The international collaboration of their R&D is very limited so far. To compensate for or augment Hong Kong's limited R&D capability (less than 50 R&D in-house staff – mostly for development work), particularly in the absorption of fundamental key technology, it is crucial to pursue collaboration with foreign R&D centers. Furthermore, the establishment of university–industry linkages or basic–applied research collaboration is absolutely crucial for the development of environmental technology. It not only encourages technology transfer from university labs to industry but also supports the dynamic interaction between basic research and more applied research, which has been shown to be key to the successful development of several environmental technologies in Japan, such as nanomaterials, photocatalyst coating, and membranes and filters for wastewater treatment. Hong Kong should create both funding and evaluation schemes to encourage the interaction between university and industry, and between basic and applied research, which is a more dynamic and continuous long-term process. Hong Kong has also been working hard to meet the world standard for IPR. However, it should also recognize the danger of the IPR regime becoming overly excessive in implementation, which may discourage technological learning as well as cumulative and interdisciplinary research necessary for the development of environmental technology. In addition, incentive measures for R&D in the private sector are needed. A recent study on the Japanese experience conducted by RIETI-METI illustrates that tax breaks for green-tech R&D can be more effective than R&D subsidies.[14]

For the demand-pull development, the HKPC began to push for environmental compliance by the Hong Kong factories currently operating in the PRD region. When one tries to improve the quality of the environment, the cost of compliance goes up accordingly. This is particularly so in the PRD, where there are over 56,000 factories operating. Thus, the key is to address the problems in the region directly both through air pollution control and the improvement of energy efficiency while also considering the costs of implementation. HKPC has already introduced tough environmental management standards, such as ISO14001, and EU Environmental Directives on electronic and electrical products in the contracted factories in the region. Furthermore, the Hong Kong Environmental Industry Association can be a good platform for the industrial sector and the government to work together. Through this public–private partnership, pushing for the implementation of tougher environmental standards will inevitably lead to the expansion of the market for Hong Kong's environmental technology in the coming years.

Finally, the development of human capital is also a key policy initiative. The government of Hong Kong can clearly help to promote the "green productivity" movement by supporting various training and educational programs and introducing cleaner production technologies to energy saving manufacturing systems to industries such as construction, textiles, toys, and electronics. As well as the supporting of Hong Kong's R&D network of global innovation in environmental technology through building up the innovation capabilities through a stable funding system, bringing in top researchers and engineers to the region for consultation and training is also important. For example, HKPC has established various training programs and study missions over the last 25 years. However, in order to introduce environmental technology in the region more effectively, each project must be uniquely tailored for individual factories. Thus, human resource must be constantly available. Furthermore, it is apparent that a pool of young talent and well-qualified students is necessary to maintain and improve the innovation capabilities of Hong Kong in any areas of technological advance. As in China, Hong Kong should also take advantage of the effect created by "brain circulation" to help raise the level of research. One of the first objectives of innovation policy for Hong Kong in today's global knowledge-based economy should be to attract not only students from both Hong Kong and Mainland China, but also from the global talent pool beyond East Asia.

Notes

1. For China's Mid- and Long-Term Plan for "green-tech" development strategy, please see, http://www.ndrc.gov.cn/fzgh/ghwb/115zxgh/PO20070930491947302047.pdf.
2. Japan 'Science and Technology' Agency, *Kagakugijyutsu Kenkyukaihatsu no Kokusaihikaku*, 2008. JST Report.
3. INSEAD and CII, *Global Innovation Index*, 2008.
4. The Chinese government's commitment to technological development leading to its sustainable development can also be found in the Mid- and Long-Term S&T Development Plan, 2007 please see, http://www.ndrc.gov.cn/fzgh/ghwb/115zxgh/PO20070930491947302047.pdf.
5. WEEE stands for "Waste Electrical and Electronic Equipment." RoHS stands for "Restriction of Use of Certain Hazardous Substances in Electrical and Electronic Equipment," and EuP stands for "Framework for the Setting of Eco-design Requirements for Energy-using Products."
6. Based on an interview, 2008 with Mr Anthony Ma, HKPC.
7. Japan Science Technology Agency (2008).
8. Based on the hearing at HKSAR in 2008.
9. Based on the hearing at NAMI in 2008.
10. Hearing at the Innovation and Technology Commission.
11. Based on an interview with Professor Jimmy Yu at the Chinese University of Hong Kong 2008.
12. Japan Science and Technology Agency, *Kagakugijyutsu Kenkyukaihatsu no Kokusaihikaku*, 2008. JST Report.
13. National Institute of Science and Technology Policy (Japan), *Study of the World's Top Class Research Centers in Europe*, NISTEP Report No. 112, March 2008.
14. K. Motohashi, Research Institute of Economy Trade and Industry (RIETI) Discussion Paper, 2008.

References

Global Innovation Index 2008–2009, INSEAD, 2009.
Innovation and Technology Commission, *Consultation Paper: New Strategy of Innovation and Technology Development*, June 2004.
Innovation and Technology Commission, *Laying the Foundation and Forging Ahead*, May 2007.
Japan Science Agency, *Kagakugijyutsu Kenkyukaihatsu no Kokusaihikaku*, 2008. JST Report.
Lau Pui King, *Shenzhen-Hong Kong Technology Industry Cooperation*, TDC Research Paper, 2007.
The State Council of China, *Medium- and Long-term National Plan for Science and Technology Development 2006–2020* (Policy document originally in Chinese).
National Institute of Science and Technology Policy (Japan), *Study of the World's Top Class Research Centers in Europe*, NISTEP Report No. 112, March 2008.
Woodrow Wilson International Center for Scholars, *China Environment Series*, Issue 9, 2007.

Hong Kong government sources:

Hong Kong Productivity Council
Department of Environmental Protection
Energy Division of the Environment Bureau
Sustainable Development Division of the Environmental Bureau
Innovation and Technology Commission.

Other sources:

Professor Jimmy Yu, the Chinese University of Hong Kong
Nanotechnology and Advanced Materials Institute (NAMI), Hong Kong University of Science and Technology
Auto Parts and Accessory Systems (APAS)–R&D Center, Hong Kong
Hong Kong Science and Technology Park
Nature China.

Index